Environmental Treatment Technologies for Municipal, Industrial and Medical Wastes

Environmental Treatment Technologies for Municipal, Industrial and Medical Wastes

Remedial Scope and Efficacy

Second Edition

Subijoy Dutta

CRC Press
Taylor & Francis Group
Boca Raton London New York

CRC Press is an imprint of the
Taylor & Francis Group, an **informa** business

CRC Press/Balkema is an imprint of the Taylor & Francis Group, an informa business

Library of Congress Cataloging-in-Publication Data

Names: Dutta, Subijoy, author.
Title: Environmental treatment technologies for municipal, industrial and
medical wastes : remedial scope and efficacy / Subijoy Dutta.
Description: 2nd edition. | Boca Raton : CRC Press, [2022] | Includes bibliographical references and index.
Identifiers: LCCN 2021021500 | ISBN 9780367435509 (Hbk) | ISBN 9781032058214 (Pbk) | ISBN 9781003004066 (eBook)
Subjects: LCSH: Factory and trade waste.
Classification: LCC TD897 .D88 2022 | DDC 628.4--dc23
LC record available at https://lccn.loc.gov/2021021500

Published by: CRC Press/Balkema
Schipholweg 107C, 2316 XC Leiden, The Netherlands
e-mail: Pub.NL@taylorandfrancis.com
www.crcpress.com – www.taylorandfrancis.com

ISBN: 9780367435509 (Hbk)
ISBN: 9781032058214 (Pbk)
ISBN: 9781003004066 (eBook)

DOI: 10.1201/9781003004066

Dedication

This book is dedicated to the beloved memory of My Father, Subinoy Kumar Dutta (1916-1990) and Mother, Santwana Dutta (1924-1996)

Contents

Foreword

Throughout history, humans have developed technologies to meet our needs for food, energy, water, and medicines as well as specific products that make our industries efficient and our daily lives more convenient. These activities often create by-products that are discarded or which escape into the environment. Some by-products—food waste, say—degrade naturally and pose little direct risk to people or ecosystems. But other by-products are not so benign.

As an example, the Housatonic River in Massachusetts is contaminated with polychlorinated biphenyls (PCBs) discharged by a General Electric plant in Pittsfield. The PCBs were integral to power transformers that were serviced in the plant for more than four decades of the mid-twentieth century. The contamination runs many miles downriver, and it also affects lands associated with the former plant. Left alone, the PCBs would not degrade for hundreds of years, and so the federal government has mandated removal of the contaminated soils and sediments. PCBs were banned in 1979, but the clean-up efforts along the river continue even to this day.

Other examples are only too easy to find. Starting in the 1850's, hydraulic mines in Californiaseparated gold from raw ore by amalgamating it with mercury. An estimated 10%–30% of that mercury escaped into river sediments, and it still poses health risks 150 years later. A tannery that once operated in Woburn, Massachusetts, not from my home, left soils contaminated with hexavalent chromium. The Hanford Nuclear Reservation in Washington leaks radionuclides into ground water near the Columbia River, mostly by-products of projects from more than 50 years ago. More recently, attention has turned toward per- and polyfluoroalkyl substances (PFAS)—sometimes called "forever chemicals" for their resistance to environmental degradation—which now appear in drinking water supplies around the world.

Cleanup of these and other substances can be both complex and costly, and the results must achieve specific standards set by environmental regulations. A vast array of technologies has been developed for many categories of hazard, whether in soil or water, whether organic compounds or heavy metals, and whether by-products of manufacturing, mining, or military ordnance. The wide variety of contaminants, settings, and solutions would challenge any expert's knowledge.

In this book, Subijoy Dutta guides the reader through these many complexities. Dutta served at the US Environmental Protection Agency (EPA) for 20 years as a senior engineer, where his work addressed soils, sediments, landfills, and pesticides. He has also worked with the Central Pollution Control Board in New Delhi on medical wastes and landfills, and he has been recognized by the EPA for his work on cleaning the Yamuna River in India.

The book begins by introducing the regulatory background. The initial step of developing a site management plan is laid out, including the need to prevent cross-contamination. For example, volatile organic compounds may escape and contribute to photochemical smog, or greenhouse gas emissions may compound climate change. Groundwater treatment is discussed next. Containment—to simply hold contaminants in place—is described, followed by methods of soil washing and thermal treatment of soils. There follow chapters on vapor extraction, bioremediation, incineration, and other approaches. Very usefully, a full chapter is devoted to case studies, detailing the solutions to a number of real-life situations. These include soil and water contaminated by lead, chromium, organic solvents, pesticides, and petroleum derivatives. The book then turns to methods for monitoring and control of contaminants, and finally discusses issues specific to medical wastes.

This comprehensive work will serve working engineers, government regulators, and environmental stewards. The reader will be solidly grounded in a wide range of solutions for environmental remediation. And such solutions will surely continue to be needed for a long time to come.

John H. Lienhard V, PhD, PE
Director, Abdul Latif Jameel Water and Food Systems Lab
Massachusetts Institute of Technology
November 2020

Preface

The motivation for writing this book comes from a much-felt need for safe implementation of remediation technologies with adequate controls against migration of contaminants to ambient air, soil, or water. During my visits to several contaminated sites in the US and other countries, I have noticed the lack of focus toward migration of contaminants to the ambient environment during cleanup activities. The cleanup operations are generally focused on achieving the cleanup goals without a full mass balance type analyses of the contaminant/s removed. I noticed more conspicuously at different sites that during cleanup activities, there were lack of attention towards these unaccounted air emissions or migration, referred to as cross-media transfer, of contaminants. This poses the problem of transferring contaminants from one media to another while assuming proper cleanup of the contaminated media. To that end, the fugitive emission of VOCs adds to the greenhouse gas in the Earth's lower atmosphere leading to global warming and climate change. This is a serious concern at the present time, and efforts to control these emissions by looking at possible escape points in various steps of environmental treatment technology applications are elaborated and emphasized all throughout this book.

The book specifically lists the best management practices (BMPs) and options for controlling these emissions and migrations through the escape points during technology implementations. The BMPs provide guidance on how to design and conduct remediation activities with minimal emissions or leaks of contaminants to air or water. At most of the contaminated sites, during remediation activities, there are some common steps for implementing the cleanup technologies. Regardless of the selected technology, these activities generally fall within one of the four major remedial stages – Site Preparation and Staging, Pre-Treatment, Treatment, and Post-Treatment/Residual Management. Appropriate BMPs to control emissions or migrations during each of these stages are listed in this book. During the late 1990s, I worked with a team of professionals from the US Environmental Protection Agency (EPA), a few States, and industries to develop the *BMPs for Soil Treatment Technologies*. Based on the observations and performances of these BMPs, I have updated the BMPs in this book for proper field applications and use at various sites. I hope they are put to practice for controlling and minimizing emissions and migrations from waste sites.

Realizing the importance to control the greenhouse gas emissions and the related impacts due to climate change, this book focuses on cross-media transfer of contaminants and their control using the best management practices/options (BMPs). Proper emission control of volatile organic compounds at the waste sites during treatment, storage and disposal activities safeguards health of local residents and minimizes global warming and related disasters due to climate change.

Most of you know about the severe pollution in the Yamuna River in India, the Pasig River in the Philippines, the Thames River in the UK, and various other polluted rivers across the world. The pollution loads and statistics are alarming in some cases. In April 1992, I began to look into possible remedial options for the Yamuna near New Delhi, India, and came up with a few ideas including constructed wetland systems and other low-cost treatment systems. I interacted with the New Delhi authorities to start a low-cost sewage treatment system on a demonstration basis. The authorities were constantly changing hands and the project didn't get started there. However, I looked into other authorities and completed a demonstration project involving an innovative wastewater treatment system near Hyderabad, India. I also got involved with a few superfund site remediation designs and implementations in the US. After being actively involved in the field for the past 28 years, I tried to provide some guidelines in this book on the site remediation process beginning from preliminary assessment and site investigations to remedial design, remedial actions, site closure, and post-closure activities, which should be applicable to both developed and developing countries.

The purpose of this book is to dissipate information on currently applicable and available technologies to the Americas, European Countries, Indian subcontinent, Southeast Asian countries, Australia, New Zealand, and other countries. I feel that this book would provide useful tools and options for improving the awareness and control of environmental pollution in both developing and developed countries. This book is intended for almost anyone who intends to enhance their knowledge in environmental pollution control technologies and waste management. After completion of site assessments, this book should provide valuable information in selecting the most environmentally safe and effective remedies and in conducting the remedial design and implementation for individual sites.

This book provides detailed information on various waste treatment technologies for industrial, municipal, and medical wastes. All available waste treatment technologies have been grouped into several technology groups, which contain similar technologies. The following treatment technology groups are covered in separate chapters of this book:

- Groundwater treatment;
- Containment technology;
- Soil washing;
- Thermal treatment;
- Vapor extraction;
- Bioremediation;
- Incineration;
- Other physical/chemical treatments;
- Monitoring and control technologies;
- Medical waste treatment technologies.

Chapter 1 of the book provides introduction and a general regulatory background of various countries. It also covers waste definition and characterization. The primary hierarchy of environmental regulation involving recordkeeping, manifest preparation, storage, and disposal is also covered in this chapter. Site inspection, assessment, remedy selection, implementation, and post-closure are touched upon at the end of this chapter. Chapter 2 provides details on the site remediation process. The cleanup process is elaborated through a flowchart, which should provide a very clear, step-by-step process of remediation. This chapter covers preliminary assessment, remedial investigation, feasibility study, remedial design, remedial action, and closure and post-closure activities.

The climate change aspects of greenhouse gas emissions during the remedial process are also included in Chapter 2 with a brief synopsis of findings on the significance of climate change by the European Commission and the National Aeronautics and Space Administration (NASA). A figure providing supporting data from NASA from 1850 to 2020 on solar irradiance and temperature of the Earth's lower atmosphere is included. The high levels of volatile organic contaminant (VOC) emissions from the waste piles during transportation, storage, and actual remediation activities, without proper BMPs or control, contribute toward the climate change by adding greenhouse gases to the atmosphere. Realizing the importance to control these greenhouse gas emissions, this book is focused on identification of emission points during the complete waste remediation process and control of such cross-media transfer of VOCs using the best management practices/options (BMPs) specified in this book.

Chapter 3 covers groundwater treatment technologies as the primary method of treating contaminated groundwater in site remediation process. Contaminants from municipal and industrial wastes often migrate and pollute the groundwater. This is a common issue in many Superfund sites in the US. While the rest of the book covers primarily the municipal or industrial wastes containing soils, sediments, and sludges, this chapter provides detailed reviews of various groundwater treatment technologies used for site remediation. Various factors such as long-term effectiveness and permanence, reduction of toxicity, mobility or volume, short-term effectiveness, implementability, cost, acceptance by the state and local authorities, and community acceptance have been evaluated and analyzed to determine the suitability of a particular technology at a specific site. The efficacy and applicability of a variety of groundwater treatment technologies have been tabulated at the end of this chapter.

Chapter 4 addresses the containment treatment technology group. All different variations of containment technologies, such as storage containers, drums, tanks, impoundments, slurry walls, landfill covers and liners, cement-based barriers, and geomembrane barriers, have been covered in this chapter. Control practices to prevent migration of contaminants to ambient environment during remedy implementation have been addressed in this chapter for various containment technology implementations.

Soil washing and thermal treatment have been addressed in Chapters 5 and 6 of this book, respectively. The definition and scope of these technologies, description of the technologies, cross-media transfer potential during the implementation of various technologies in these technology groups, and best management options to control the cross-media transfer of contaminants have all been covered in these chapters.

Similarly, Chapters 7, 8, 9, and 10 address vapor extraction, bioremediation, incineration, and other physical/chemical technology groups, respectively. There are a number of technologies listed under each technology group in the respective chapters, and descriptions of some of those technologies are provided. The treatment technology details along with the likelihood of cross-media transfer during the complete remediation process and their control aspects have been addressed in these chapters.

Chapter 11 presents a unique assimilation of ten case studies. All of the ten case studies are taken from various regions and states within the US. It involves field applications of various treatment technologies, such as soil washing/soil leaching to treat metal contaminated soils; in situ chemical stabilization at a petroleum refinery site; ex situ stabilization and disposal of lead contaminated soil at a battery manufacturing facility; size separation/soil washing of metals; soil vapor extraction at an electronic component manufacturing facility; on-site containment of hexavalent chromium contaminated soils at a former chromium plant; bioremediation of explosives contaminated soils; air sparging of groundwater combined with vapor extraction of soil; thermal desorption at a blending and packaging site; and in situ bioremediation using molasses injection. All of the remediation steps including site preparation/staging, pre-treatment, treatment, and post-treatment activities for each of the ten case studies are covered. The focus of these case studies is on the use of best management practices during the implementation of various treatment technologies to prevent air emissions or migration of contaminants.

Chapter 12 covers monitoring and control of air emissions or migration of contaminants during common activities, which are involved in any site cleanup. It provides a general guideline of practices for controlling migration of contaminants during these common cleanup activities.

Monitoring and control of cross-media transfer of contaminants during the implementation of specific site cleanup technologies are addressed in Chapter 13. Many control technologies have been tabulated and methodically presented in this chapter to help the reader with quick and easy identification of effective control technologies to meet their specific needs during remedial actions. This chapter also covers remote sensing technologies for monitoring and control of site remediation activities. It includes an example of monitoring and control of a waste site in a remote area, which was more precise and economic with the use of remote sensing technologies.

Chapter 14 addresses medical waste treatment and disposal options. This chapter presents available treatment options for medical waste and provides a synopsis of medical waste management. Information on safe management of infectious and other routine waste from medical facilities is incorporated in this chapter. It identifies necessary tools and techniques for safe management of medical wastes to meet the needs of medical professionals from different parts of the world. This chapter begins with medical waste identification and profile, looks into the categorization of medical waste, and various waste minimization options. It covers the impending issue of proper disinfection and disposal of COVID-19 related waste (C19-waste) pertaining to the current pandemic. It provides stepwise guidance on a safe and manageable way for hospitals and other health care facilities to dispose of their C19-waste. These globally applicable guidance are based upon recommended practices by US Centers for Disease Control (CDC), European Commission (EC), and the World Health Organization (WHO). Medical waste treatment technologies are covered in depth followed by an evaluation and selection of treatment options at the end of this chapter. The air emission issues for incineration, commonly used as an off-site treatment for medical waste, are covered in Chapter 9, and the related impact on climate change is covered in Chapter 2.

A list of references, which should be useful for the readers to get further information on a specific area of their interest, is provided at the end of each chapter. I hope that readers from many parts of the globe would benefit from the information provided in this book. The more we prevent the contaminants from migrating into our ambient environment, the better will be the environmental health of our world resulting in a healthier life with cleaner air, soil, and water.

Subijoy Dutta

Acknowledgments

My sincere thanks go to Dr. John H. Lienhard, professor, Massachusetts Institute of Technology for reviewing the entire book and providing valuable comments in his Foreword for the book. My deep appreciation and thanks go to Dr. William E. Roper and Dr. Darrell Cornell for providing encouragement to me in writing this book. Periodic feedback by Dr. Roper was of great value. I am thankful to Dr. Kumar Kanti Das, F.A.C.S, F.R.C.S, Silchar, India for his review of the medical waste Chapter of the book and providing valuable comments. My thanks and appreciation go to Dr. Shyam Mathur, senior chemist and my ex-colleague from the US Environmental Protection Agency, for his valuable thoughts and discussion on some of the new remediation areas. I want to express my thanks to a number of others in the US, Canada, India, Indonesia and Philippines who participated in my environmental training classes and provided valuable input on local practicability of technologies. I am also thankful to the S&M Engineering team – Sucharit Dutta, Phalguni Bhattacharyya and others – in India for their support in providing valuable information from the field.

My deep appreciation and thanks go to our sons, Dr. Sumit Dutta and Sanjit Dutta for providing great support in reviewing a number of figures in this book. Their invaluable interactions and inputs were instrumental in selecting the final cover design of the book. My sincere thanks go to my wife, Urmi Dutta, who provided a great support to me while I was writing this book. There are many others who have encouraged, helped, and supported me in different ways – such as Daryl Jackson in Poteau Oklahoma, Dr. Raj Rajaram and Dr. Prakasam Tata in Chicago area, and the list goes on. It is not possible to list everyone's name here, but I'd like to convey my appreciation to everyone who worked, helped, and supported in any manner toward completion of this book and its final production.

Author

Mr. Subijoy Dutta is the founding director of *Rivers of the World Foundation* and sole proprietor/Director of S&M Engineering, LLC (http://snmengineering.com).

He authored and co-authored the following two books:

- *Environmental Treatment Technologies for Hazardous and Medical Wastes – Remedial Scope and Efficacy*, published in March 2002 by Tata McGraw Hill Publishing Company.
- *Sustainable Mining Practices – A Global Perspective*, published in July 2005 by Balkema Publishers (now a subsidiary of Taylor & Francis Group, UK).

Mr. Dutta received his BS in Mechanical Engineering from the Assam Engineering College, Guwahati, India in 1972. He completed his first MS in Mechanical Engineering from the University of Oklahoma, Norman, Oklahoma, USA in 1981. In 1984, he finished a second MS in Petroleum/Geological Engineering from the same Oklahoma University.

In 1992, he stepped forward to provide help towards cleaning the Yamuna River. He collected an inordinate amount of data, and met with the Delhi authorities and offered to provide assistance. He formed the Yamuna Foundation in 2000 and had many local volunteers in Agra and Delhi area. He later founded the Rivers of the World Foundation in 2007 along with a few esteemed professionals in the US.

Mr. Dutta worked for the US Environmental Protection Agency (EPA) in Washington, DC for 20 years as a Senior Engineer. He has assisted the Central Pollution Board, New Delhi, in their development of regulations and guidance concerning medical waste management and landfills.

The following is a list of his major project experience:

- He designed an innovative well interception plan to successfully kill the 1.4m gal/day Deepwater Horizon oil spill (Macondo Well) in August 2010. He used the bathymetric profile of the seabed to locate the interceptor well.
- As a member of the several workgroups, while with US EPA, he was involved in revising many regulatory standards and guidance, such as

- State Management Plans (SMP) for the Office of Pesticide Programs,
- Total Maximum Daily Load (TMDL), hazardous waste identification rule,
- Soils and sediments standards, the wetland guidance document, and several fact sheets.

- Conducted an EPA soil leaching study in 1993 using contaminated soil samples from various sites representing the spectrum of soils and climatic conditions all across the continental US.
- Designed modified cover systems for landfills and has supervised installation of landfill covers for several hazardous waste sites.
- Provided waste management training to 26 engineers with the Tamil Nadu Pollution Control Board in 1998.
- He assisted the Central Pollution Control Board, New Delhi, India in their development of regulations and guidance concerning medical waste management and landfills by providing voluntary review and comments.
- Prepared reclamation plans for several abandoned mine sites in Eastern Oklahoma.
- Recently developed a watershed management plan by using GIS and remote sensing methodologies for a very fast-growing county in Virginia.
- Used/explored remote sensing technologies for various applications, such as Oil and Gas Pipeline Condition Assessment, analyzing various imageries and signatures for oil and gas explorations, and pipeline security assessments.
- He received the "Unsung Hero" award from the EPA Administrator, Carol Browner for his work on the Yamuna River protection and restoration.

ACRONYMS

AEC	Area of Environmental Concern
ALARA	As Low As Reasonably Achievable
APCD	Air Pollution Control Device
ASME	American Society of Mechanical Engineers
ASTM	American Society of Testing and Materials
ATP	Anaerobic Thermal Process
BACT	Best Available Control Technology
BDAT	Best Demonstrated Available Technology
BMPs	Best Management Practice(s)
BOD	Biochemical Oxygen Demand
BPCT	Best Practicable Control Technology
BPT	Best Practicable Technology
BTEX	Benzene, Toluene, Ethylbenzene, and Xylene
C19-waste	COVID-19 Related Waste
CAMU	Corrective Action Management Unit
CAO	Corrective Action Order
CAP	Corrective Action Plan
CDC	US Center for Disease Control
CEMS	Continuous Emission Monitoring System

CERCLA	Comprehensive Environmental Response, Compensation, and Liability Act of 1980
CFM	Cubic Feet Per Minute
CFR	Code of Federal Regulations
CFS	Cubic Feet Per Second
CMI	Corrective Measure Implementation
CMIPP	Corrective Measure Implementation Program Plan
COD	Chemical Oxygen Demand
COVID-19	novel Coronavirus Disease - 2019
CSFS	Contaminated Soil Feed Stockpiles
CWA	Clean Water Act
DEP	Department of Environmental Protection
DM	Dust Monitor
DNAPL	Dense Non-Aqueous Phase Liquid
DNR	Department of Natural Resources
DoD	US Department of Defense
DoE	US Department of Energy
EC	European Commission
EIA	Environmental Impact Assessment
EIS	Environmental Impact Statement
EPA	US Environmental Protection Agency
ESP	Electrostatic Precipitators
FACA	Federal Advisory Committee Act
FCC	Federal Communications Commission
FFA	Federal Facility Agreement
FID	Flame Ionization Detector
FR	Federal Register
GAC	Granular Activated Carbon
GCLs	Geosynthetic Clay Liners
GC/MS	Gas Chromatography/Mass Spectrometry
GIS	Geographic Information System
GPD	Gallons Per Day
GW	Groundwater
HCW	Health Care Waste
HEPA	High-Efficiency Particulate Air
HMIWI	Hospital/Medical/Infectious Waste Incinerator
HWIR	Hazardous Waste Identification Rule
IRP	Installation Restoration Program
ISM	Industrial, Scientific, and Medical
ISV	In Situ Vitrification
LDPE	Low-Density Poly Ethylene
LDR	Land Disposal Restrictions
LIDAR	Light Detection and Ranging
LLRW	Low-Level Radioactive Waste
LNAPL	Light Non-Aqueous Phase Liquid
MPCA	Minnesota Pollution Control Agency
NIOSH	National Institute of Occupational Safety and Health

NoC	Notice of Construction
NPDES	National Pollutant Discharge Elimination System
O&M	Operation & Maintenance
OD	Outside Diameter
OPC	Other Physical/Chemical Treatment
OSHA	Occupational Safety and Health Administration
OVA	Organic Vapor Analyzer
PA	Preliminary Assessment
PASP	Perimeter Air Sampling Program
PCE	Perchloroethylene
PIC	Products of Incomplete Combustion
PID	Photoionization Detector
PM	Particulate Matter
PM-10	Particulate Matters Less than 10 microns in diameter
POHC	Principal Organic Hazardous Constituent
PPB	Parts Per Billion (μg/L
PPE	Personal Protective Equipment
PPM	Parts Per Million (mg/L)
PPMV	Parts Per Million by Volume
PRP	Potentially Responsible Party
PVC	Polyvinyl Chloride
RA	Remedial Action
RCRA	Resource Conservation and Recovery Act
RD	Remedial Design
RF	Radio Frequency
RFI	RCRA Facility Investigation
RI	Remedial Investigation
RI/FS	Remedial Investigation/Feasibility Study
RoD	Record of Decision (used in the Superfund Program of EPA)
SARA	Superfund Amendments and Reauthorization Act of 1986
SDWA	Safe Drinking Water Act
SHSP	Site Health and Safety Plan
SI	Site Inspection
SIS	Solidification/Stabilization
SITE	Superfund Innovative Technology Evaluation
SPCC	Spill Prevention, Containment, and Countermeasure (CWA)
SPLP	Synthetic Precipitation Leaching Procedure (EPA Method 1312)
SU	Standard Unit
SVE	Soil Vapor Extraction
SVOC	Semi-Volatile Organic Compound
SWDA	Solid Waste Disposal Act
TAP	Toxic Air Pollutants
TCE	Trichloroethylene
TCLP	Toxicity Characteristic Leaching Procedure (EPA Method 1311)
TDU	Thermal Desorption Unit
TEQ	Toxic Equivalents
TNT	Trinitrotoluene

TOC	Total Organic Compounds
TPH	Total Petroleum Hydrocarbons
TRPH	Total Recoverable Petroleum Hydrocarbons
TRU	Transuranic Waste
TSDF	Treatment, Storage, and Disposal Facility
TSP	Triple Superphosphate
TSS	Total Suspended Solids
TWA	Time Weighted Average
ULPA	Ultra Low Penetration Air Filter
US EPA	US Environmental Protection Agency
UST	Underground Storage Tank
UV	Ultraviolet radiation
UXO	Unexploded Ordnance
VOC	Volatile Organic Compound
WHO	World Health Organization
WIPP	Waste Isolation Pilot Plant
XRF	X-Ray Fluorescence

Chapter 1

Introduction

1.1 REGULATORY BACKGROUND

In determining the extent and refinement of the treatment technologies to be deployed for environmental remediation of municipal, industrial, and medical wastes, the existing regulatory requirements are generally used as the primary factors. These requirements vary enormously from country to country, state to state within a country, and even between different districts within a state. The varying regulatory frameworks and the resulting requirements around the globe are beyond the scope of this book. However, environmental regulations anywhere in the world would generally fall into one of the following three major categories:

- Regulations have been developed with input from stakeholders and are being implemented with enforcement.
- Regulations have been developed for meeting ideal conditions and are not being implemented with proper enforcement.
- Regulations are in the process of being developed, and no clear directions exist for the people/industries for safe management of their hazardous or infectious wastes.

Most of the developed countries fall into the first category. Some of the developing countries fall into the second category, and some of them fall into the third category as they do not have any environmental policies or regulations. Thus, all developing countries fall into a mixed bag of the second and third categories.

The primary hierarchy of environmental regulations for waste management generally involves:

i. Waste identification and characterization
ii. Recordkeeping and manifestations
iii. Safe storage and disposal practices
iv. Inspection and assessments of hazardous waste sites
v. Remedy selection and design
vi. Remedy implementation and site closure
vii. Post-closure and site restoration/reuse

DOI: 10.1201/9781003004066-1

1.1.1 Waste identification and characterization

When someone is handling municipal or industrial waste material, the most obvious question he/she would have is whether the material is a non-hazardous waste or a toxic/harmful or hazardous waste. A waste is considered to be a hazardous waste if it exhibits any of the following four characteristics:

- *Ignitability*: To explain in a very simple way, a waste is considered to be ignitable if a representative sample of the waste is capable, under standard temperature and pressure, of causing fire through friction, absorption of moisture or spontaneous chemical changes and, when ignited, burns so vigorously and persistently that it creates a hazard.
- *Corrosivity*: a solid waste is considered to exhibit corrosive characteristics if a representative sample of the waste is aqueous in nature and has a pH less than or equal to 2 or greater than or equal to 12.5 or it corrodes steel at a rate greater than 6.35 mm (0.250 inch) per year at a test temperature of 55°C (130°F) as determined by standard test methods.
- *Reactivity*: a reactive waste characteristic is exhibited if a representative sample of the waste is normally unstable and readily undergoes violent changes such as reacting violently with water, forming potentially explosive mixtures with water, when mixed with water or a cyanide or sulfide bearing waste exposed to a very low (2.0) or high (12.5) pH generating toxic gases/fumes in a quantity sufficient to present a danger to human health or the environment. A waste may also be considered reactive if the representative sample is capable of detonation or explosive decomposition or reaction at standard temperature and pressure.
- *Toxicity*: a solid waste exhibits the characteristics of toxicity if a representative sample from the waste contains toxic contaminant at a concentration sufficient to present danger to human health or the environment. The United States Environmental Protection Agency (USEPA) characterizes a waste to be toxic if extract (by the Toxicity Characteristic Leachate Procedure, Method 1311 or equivalent) from a representative sample of the waste is found to contain any of the following primary contaminants at levels above the maximum concentration level (per USEPA) specified in Table 1.1. Readers should note that the levels specified in this table are the deterministic levels for characteristics of toxicity only and are generally higher than the regulatory trigger levels in some cases.

Once the waste is identified or characterized to be a hazardous waste, the next question that comes to mind is how hazardous is the waste that is being handled? Since different contaminants pose quite different levels of risk depending primarily upon their concentration levels, it is imperative that a thorough analysis of the waste is first conducted and the contaminants of concern are identified. Most of the countries under category 1 have regulations and/or guidance that specifies the sampling and analysis techniques. Most developing and underdeveloped countries may not have specific sampling and analysis techniques. Suggested literature for sampling and analysis includes: Data Quality Objectives Process for Superfund – Interim Final Guidance (USEPA, 1993); Data Quality Objectives for Response Activities: Development Process (USEPA, 1987); and Quality Assurance Project Plan for Characterization Sampling and Treatment Tests Conducted for the Contaminated Soil and Debris Program (USEPA, 1990a). A few additional references are also furnished at the end of this chapter (Englund and Heravi, 1994; Hahn and Meeker, 1991; Journel, 1988; USEPA, 1986, 1990b, 1996a; Mason, 1992; Neptune et al., 1990; Patil et al., 1994; Pitard, 1993). Similarly, analytical methods for most of

Table 1.1 Maximum concentration of contaminants for the toxicity characteristics

Serial No.	Contaminant	Chemical abstract service (CAS) no.	Regulatory level[a] (mg/L)	Serial No.	Contaminant	Chemical abstract service (CAS) no.	Regula-tory level[a] (mg/L)
1.	Arsenic	7440-38-2	5.0	20.	Hexachlorobenzene	118-74-1	0.13
2.	Barium	7440-39-3	100.0	21.	Hexachloro-butadiene	87-68-3	0.5
3.	Benzene	71-43-2	0.5	22.	Hexachloroethane	67-72-1	3.0
4.	Cadmium	7440-43-9	1.0	23.	Lead	7439-92-1	5.0
5.	Carbon Tetrachloride	56-23-5	0.5	24.	Lindane	58-89-9	0.4
6.	Chlordane	57-74-9	0.03	25.	Mercury	7439-97-6	0.2
7.	Chlorobenzene	108-90-7	100.0	26.	Methoxychlor	72-43-5	10.0
8.	Chloroform	67-66-3	6.0	27.	Methylethylketone	78-93-3	200.0
9.	Chromium	7440-47-3	5.0	28.	Nitrobenzene	98-95-3	2.0
10.	o-Cresol	95-48-7	200.0	29.	Pentachlorophenol	87-86-5	100.0
11.	m-Cresol	108-39-4	200.0	30.	Pyridine	110-86-1	5.0
12.	p-Cresol	106-44-5	200.0	31.	Selenium	7782-49-2	1.0
13.	2,4-D	94-75-7	10.0	32.	Silver	7440-22-4	5.0
14.	1,4-Dichlorobenzene	106-46-7	7.5	33.	Tetrachloroethylene	127-18-4	0.7
15.	1,2-Dichloroethane	107-06-2	0.5	34.	Toxaphene	8001-35-2	0.5
16.	1,1-Dichloro-ethylene	75-35-4	0.7	35.	Trichloroethylene	79-01-6	0.5
17.	2,4-Dinitrotoluene	121-14-2	0.13	36.	2,4,6-Trichlorophenol	88-06-2	2.0
18.	Endrin	72-20-8	0.02	37.	2,4,5 TP(Silvex)	93-72-1	1.0
19.	Heptachlor	76-44-8	0.008	38.	Vinyl chloride	75-01-4	0.2

[a] USEPA (2005), Intro to Hazardous Waste Identification, EPA-530-K-05-012, Sep. 2005, per 40CFR, Part 261.24 standards, Washington, DC.

the listed contaminants are provided in Test Methods for Evaluating Solid Waste, Physical/chemical Methods, SW-846, 3rd Edition, Final Update (USEPA, 1997).

1.1.2 Recordkeeping and manifestation

Recordkeeping and manifestation of the waste are also an integral part of a safe waste management practice. It helps to provide an explicit account of all the waste generated and their disposition and allows a clean and safe management of the waste. Under such management, a facility or generator generally prepares manifests of every unit of waste, such as a drum or a container, and lists the facility which would handle the waste. Before a waste container leaves for a treatment, storage, or disposal facility, the originating facility, the waste content, the transporter, and the ultimate disposition of the waste are all entered in the manifest. Many countries have adopted standards for transportation of hazardous waste. One major component of the standard is the waste manifest system. In general, a transporter of hazardous waste is required to follow certain management standards where such regulation exists. These management standards may include: (1) requirement of a signed manifest by the generator, (2) acknowledgment of receipt of the waste by the transporter, (3) ensuring accuracy of the manifest, and (4) physical verification of the waste quantity accompanying the manifest.

1.1.3 Storage and disposal

Safe storage and disposal of hazardous waste also play key roles in protecting the human health and the environment. Almost all developed countries have basic standards for storage and disposal of hazardous waste. Other countries have either developed or are in the process of developing these standards. Most of the safe storage and disposal practices rely upon the basic premise of containing the waste and preventing it from migrating to air, soil, or water. For further information on storage and disposal, readers may consult USEPA, 1989, and/or Bhide, 1990 listed under reference at the end of this chapter.

1.1.4 Site inspection and assessment

Due to improper management of hazardous waste in the past, many facilities, buildings, yards, and natural streams or other water bodies have already been polluted in most of the developed, developing, and underdeveloped countries. An environmentally safe management of those polluted or potentially polluted sites involves inspection and assessment of the site as the first step. Site inspection and assessment generally involve the following:

- Collecting background information on the site and preparing a site location map.
- Providing site description to include various information related to the site such as geographical and political boundaries, demography of the adjacent areas, vegetation, climate, economy, and other relevant factors.
- Data collection and analysis.
- Delineating the nature and extent of the problem by use of various site assessment tools and techniques.
- Writing a summary report on the findings from the site inspection and assessment.

A detailed environmental site assessment process could be time-consuming and tend to be expensive when an inordinate number of samples are required to be analyzed. As the first step in the overall remediation process, the site assessment process is critical to making appropriate remedial decisions. When site assessments are complete, they provide accurate information about the presence and distribution of contaminants, thereby facilitating cost-effective and efficient remediation. Incomplete site assessments can provide inaccurate or misleading information which can delay effective remediation, increase overall corrective action costs, and result in an increased risk to human health and the environment. By nature, there are always gaps in the information provided in site assessments. It is, therefore, not always obvious when a site assessment is complete and when the information has been accurately interpreted. As a result, a tremendous amount of data is needed to determine where contaminants are located and how best to remediate them.

Site assessments can also contribute directly to a large percentage of the overall remediation costs. Sampling equipment, sample analysis, and labor hours may cost between 10% and 50% of the total remediation costs at various waste sites. To reduce the time and expense for the site assessment process, a new concept of expedited site assessment process has been introduced in the developed countries as a way to streamline the remediation process, improve data collection, and reduce the overall cost of remediation. For further details on expedited site assessment, readers may consult USEPA 1998 and other references listed at the end of this chapter (Journel, 1988; USEPA, 1988).

Remedy selection is conventionally followed by environmental assessment of a hazardous waste site. This involves available remedial treatment technologies, options, and effectiveness of different alternatives under varying conditions (USEPA, 1991, 1995, 1996b). In the following chapters, a general site cleanup process, various environmental treatment technologies for hazardous wastes, a few remediation case studies in the US, and medical waste management are covered in detail.

The current pandemic involving the novel coronavirus disease-2019 (COVID-19) necessitated the usage of protective face masks, gloves, and various other personal protective equipment (PPE) measures. This is leading to a rapid accumulation of potentially infectious waste streams, causing a challenge due to an unpredictably high volume of COVID-19-related waste in many cities and towns in the world.

Waste management practices in these countries and many others will be seriously strained and may fail to provide safe disposal of these infectious waste, resulting in further spread of COVID-19 to common people and healthcare workers due to their exposures.

This book includes a section titled "Waste management for COVID-19 pandemic" in Chapter 14 "Treatment options for medical wastes", which provides details on the minimization, disinfection, and disposal practices of COVID-19-related waste (C19-waste). The information provided there deals with the sudden increase/overload of this infectious C19-waste based upon interim guidance from the World Health Organization (WHO, 2020) and the Centers for Disease Control (CDC, 2020).

REFERENCES

Bhide, A.D. (1990) *Regional Overview of Solid Waste Management*. WHO, Regional Office for South-East Asia, New Delhi, January.

Centers for Disease Control (CDC). (2020) What Waste Collectors and Recyclers Need to Know about COVID-19, Washington, DC, May. [Online] Available from: https://www.cdc.gov/coronavirus/

2019-ncov/community/organizations/waste-collection-recycling-workers-h.pdf [Accessed 24th September, 2020].

Englund, E.J., and N. Heravi. (1994) Phased Sampling for Soil Remediation. *Environmental and Ecological Statistics*, vol. 1, pp. 247–263.

Hahn, G.J., and W.Q. Meeker. (1991) *Statistical Intervals: A Guide for Practitioners*. Wiley & Sons, New York.

Journel, A.G. (1988) Non-parametric Geostatistics for Risk and Additional Sampling Assessment. In *Principles of Environmental Sampling*. L. Keith, Ed. American Chemical Society, Washington, DC.

Mason, B.J. (1992) *Preparation of Soil Sampling Protocols: Sampling Techniques and Strategies*. USEPA, EMSL, Las Vegas, NV. EPA/600/R-92/128.

Neptune, D., E.P. Brantly, M.J. Messner, and D.I. Michael. 1990. Quantitative Decision Making in Superfund: A Data Quality Objectives Case Study. *HMC*, vol. 3, pp. 19–27.

Patil, G.P., Gore, S.D., and C. Taillie. (1994). *Composite Sampling: A Novel Method to Accomplish Observational Economy in Environmental Studies*. Center for Statistical Ecology and Environmental Statistics, Pennsylvania State University, University Park, PA.

Pitard, F.F. (1993) *Pierre Gy's Sampling Theory and Sampling Practice: Heterogeneity, Sampling Correctness, and Statistical Process Control*. 2nd Ed. CRC Press, Boca Raton, FL.

USEPA. (1986) *Test Methods for Evaluating Solid Waste, Physical/Chemical Methods*. 3rd Ed. and its updates, Washington, DC, November.

USEPA. (1987) *Data Quality Objectives for Response Activities: Development Process*. EPA, Washington, DC. EPA540-G-87/003 or NTIS: 9355.0-7B.

USEPA. (1988) *GEO-EAS (Geostatistical Environmental Assessment Software) User's Guide*. EMSL, Las Vegas, NV. EPA 600/4-88/033.

USEPA. (1989) *Decision-Makers Guide to Solid Waste Management*. EPA, Washington, DC. EPA/530-SW-89-072, November.

USEPA. (1990a) *Geostatistics for Waste Management: A User's Manual for the GEO-PACK (Version 1.0) Geostatistical Software System*. Robert S. Kerr Environmental Research Lab, Ada, OK. EPA/600/8-90/004.

USEPA. (1990b) *Quality Assurance Project Plan for Characterization Sampling and Treatment Tests Conducted for the Contaminated Soil and Debris Program*. Office of Solid Waste, Washington, DC, November 1990.

USEPA. (1991) *A Guide: Methods for Evaluating the Attainment of Cleanup Standards For Soils and Solid Media*, OSWER, Washington, DC. Publication: 9355.4-04FS, July.

USEPA. (1993) *Data Quality Objectives Process for Superfund – Interim Final Guidance*. OSWER Publication 9355.9-01, Washington, DC. EPA 540-R-93-071; NTIS: PB94963203, September 1993.

USEPA. (1995) *Superfund Soil Screening Level Guidance*. EPA, Washington, DC. EPA 540/F_95/041; NTIS #: PB.963501.

USEPA. (1996a) *Soil Sampling Guidance: Fact Sheet*. EPA, Washington, DC. EPA/540/F-95/041; NTIS#: PB96-963501.

USEPA. (1996b) *Soil Screening Guidance: User's Guide*. 2nd Ed. EPA, Washington, DC. EPA/540/R-96/018.

USEPA. (1997) *Test Methods for Evaluating Solid Waste, Physical/chemical Methods, SW_846*. 3rd Ed. Final Update 3.

USEPA. (1998) *Expedited Site Assessment Tools for UST Sites: A Guide for Regulators*. Office of Underground Storage Tanks, Washington, DC.

USEPA. (2005) *Intro to Hazardous Waste Identification*. Solid Waste and Emergency Response, Washington, DC. EPA-530-K-05-012, September 2005, per 40CFR, Part 261.24 standards.

WHO. (2020) *Water, Sanitation, Hygiene, and Waste Management for SARS-CoV-2, the Virus That Causes COVID-1: Interim Guidance*. WHO and UNICEF, Geneva, Switzerland. Reference number: WHO/2019-nCoV/IPC_WASH/2020.4, July.

Chapter 2

Site remediation process

2.1 OVERVIEW OF THE GENERAL PROCESS

The environmental remediation process requires a thorough analysis and evaluation of contaminated areas to select the best cleanup method. The condition of the environment is constantly degrading as human activities are increasing at a fast pace, which, in turn, is generating a huge quantity of municipal and industrial wastes. Both municipal and industrial waste may contain hazardous and nonhazardous wastes. An ideal waste management plan should include initial screening and sorting of the waste collected and then determine what to recycle, reuse, treat, and dispose of out of the total volume of the municipal and industrial waste collected. A long-term and sustainable plan should carefully consider the following factors in order of priority for waste products that are frequently showing up in the waste stream:

- Preventing the waste type by promoting a more sustainable and environment-friendly product for use by the local people and/or industries.
- Recycling and reuse possibility of the waste stream.
- Biological treatment possibility of the waste.
- Waste-to-energy potential of the waste.
- Permanent disposal of the waste.

The remediation process becomes more effective through the identification and resolution of issues before the remedial process begins, else this can impact the progress on cleanup. The cleanup process generally includes requirements for:

- Identification and confirmation of contaminated site;
- Studying past activities and waste characterization;
- Determining the extent of soil, sediment, surface, and groundwater contamination;
- Taking necessary actions to abate risks to human health and the environment;
- Taking steps to prevent further migration of the contaminants from the waste site;
- Restoring or replacing affected or diminished water supplies; and
- Containing/storing contaminated soil at the site until completion of remediation.

A flowchart showing a conventional remediation process is shown in Figure 2.1. A step-by-step description of the remediation process is provided below.

DOI: 10.1201/9781003004066-2

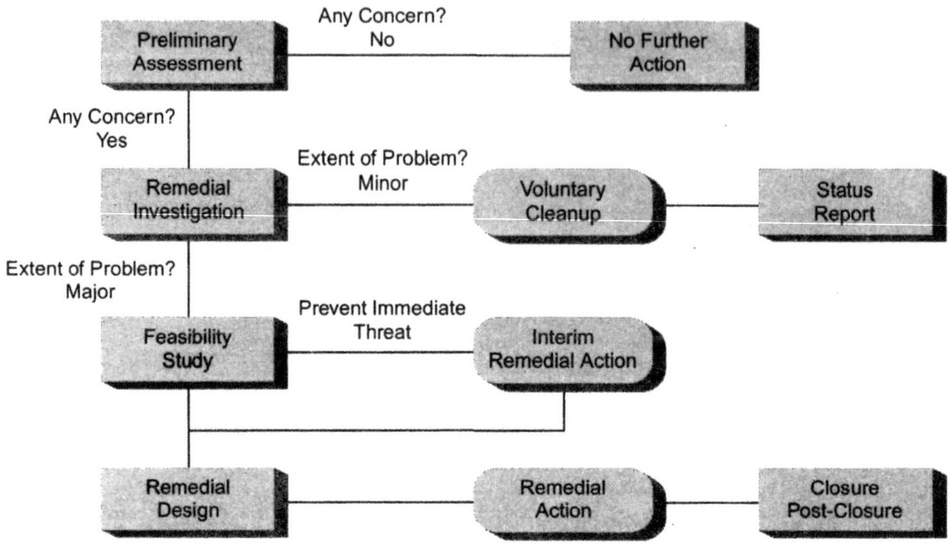

Figure 2.1 Flowchart showing a conventional remediation process (Dutta, 2002).

2.1.1 Preliminary assessment

The preliminary assessment (PA) is the very first step in the overall cleanup process. It involves gathering data to confirm/deny the existence of contamination and to analyze the potential risks to human health and the environment from a site. The assessment is generally conducted by environmental scientists or trained environmental professionals outfitted with some basic field equipment and appropriate personal protection equipment for their safety during the field investigation. Attempts should be made to keep any hazardous waste site intrusion as minimal as possible. Taking detailed pictures and field sketches of the site would enable gathering of an enormous amount of useful data. The PA either supports the need for remediation starting with a detailed remedial investigation (RI) or shows that no further action is needed. The results are generally released through a decision document or a summary report.

This step consists of a limited scope investigation performed on a potentially contaminated site. The PA investigations collect readily available information about a site and its surrounding area. The PA is designed to distinguish, based on limited data, between sites that pose little or no threat to human health and the environment and sites that may pose a threat and require further investigation. The PA also identifies sites requiring assessment for possible emergency response actions. If the PA results in a recommendation for further investigation, a site inspection (SI) is performed. Readers may refer to the publication cited under reference (USEPA, 1991) for further assistance in conducting PAs. PAs are followed by a detailed SI when the potential site is suspected to be contaminated. During SI investigations, typically environmental and waste samples are collected to determine what hazardous substances are present at a site. The investigators determine if these substances are being released to the environment and assess if they have reached nearby targets. The SI can be conducted in one stage or two. The first stage, or focused SI, tests hypotheses developed during the PA and can yield information sufficient to prepare an overall rating for the site. Different types of rating programs exist in developed countries. The commonly used methodology by USEPA's Superfund program is the hazard ranking system (HRS). It is a numerically based screening system that uses information from initial, limited investigations – the PA and the SI to assess the relative potential of sites to pose a threat to human health or the environment (USEPA, 1992). In general, the information collected to develop HRS scores is not sufficient for quantifying the extent of contamination or the remedial action (RA) for a particular site. EPA generally relies on more detailed studies in the RI/feasibility study (FS) that typically follows the remediation process (USEPA, 1988). The HRS uses a structured analysis approach to scoring sites. This approach assigns numerical values to factors that relate to risk based on conditions at the site. The factors are grouped into three categories:

1. A site has released or has the potential to release hazardous substances into the environment;
2. Characteristics of the waste (e.g., toxicity and waste quantity); and
3. People or sensitive environments (targets) affected by the release of hazardous substances.

Four pathways can be scored under the HRS:

- Groundwater migration (drinking water);
- Surface water migration (drinking water, human food chain, and sensitive environments);
- Soil exposure (resident population, nearby population, and sensitive environments); and
- Air migration (population and sensitive environments).

After scores are calculated for one or more pathways, they are combined using a root-mean-square equation to determine the overall site score. An electronic scoring system can be used to do the scoring calculations. If all pathway scores are low, the site score is low. However, the site score can be relatively high even if only one pathway score is high. This is an important requirement for HRS scoring because some extremely dangerous sites pose threats through only one pathway. Readers may refer to the publication cited under reference (USEPA, 1992) for further information on HRS. However, a regulatory agency can develop their own rating methodology depending on the local applicability of the environmental hazards and risks.

2.1.2 Remedial investigations

This is the next step after PA when the result of PA and SI indicates that a site poses potential threat or hazard to human health and the environment. In general, the RIs and the FS go hand in hand in this step. The RI serves as the mechanism for collecting data to:

- Characterize site conditions;
- Determine the nature of the waste;
- Assess risk to human health and the environment; and
- Conduct treatability testing to evaluate the potential performance and cost of the treatment technologies that are being considered.

An FS is a mechanism for the development, screening, and detailed evaluation of alternative RAs. The RI and FS are conducted concurrently – data collected in the RI influence the development of remedial alternatives in the FS, which, in turn, affect the data needs and scope of treatability studies and additional field investigations. This phased approach encourages the continual scoping of the site characterization effort, which minimizes the collection of unnecessary data and maximizes data quality.

The RI/FS process generally includes the following phases.

2.1.2.1 Scoping

Scoping is the initial planning phase of the RI/FS process. Many of the planning steps developed and outlined during scoping are continued and refined in later phases of the RI/FS. Scoping activities typically begin with the collection of existing site data; including data from previous investigations such as the PA and SI. Based on this information, site management planning is undertaken to:

- Preliminarily identify boundaries of the study area;
- Identify likely RA objectives and whether interim actions may be necessary or appropriate; and
- Establish whether the site may best be remedied as one or several different units, generally termed as operable units.

Once an overall management strategy is agreed upon, the RI/FS for a specific project or the site as a whole is planned. Typical scoping activities include:

- Initiating the identification and discussion of potential applicable or relevant and appropriate requirements (ARARs) with other supporting state, federal, or local agencies. The applicable or relevant and appropriate requirements refer to any statute or regulation that pertains to the protection of human health and the environment in addressing specific conditions or use of a particular cleanup technology at a site.
- Determining the types of decisions to be made and identifying the data and other information needed to support these decisions.
- Assembling a technical advisory committee to assist in activities, serve as a review board for important deliverables, and monitor progress during the study.
- Preparing the work plan; sampling and analysis plan; health and safety plan; and community relations plan.

2.1.2.2 Site characterization

Field sampling and laboratory analyses are initiated during the site characterization phase. A preliminary site characterization summary is prepared to provide the lead agency with information on the site early in the process before preparation of the full RI report. This summary is useful in determining the feasibility of potential technologies and in assisting both the lead and support agencies with the initial identification of the ARARs. The summary can also be sent to other agencies and stakeholders to assist them in performing their evaluation and assessments. A baseline risk assessment is developed to identify the existing or potential risks that may be posed to human health and the environment by the site. Because this assessment identifies the primary health and environmental threats at the site, it also provides valuable input to the development and evaluation of alternatives during the FS.

2.1.2.3 *Development and screening of alternatives*

The development of alternatives usually begins during scoping, when likely response scenarios may first be identified. The development of alternatives requires:

- Identifying RA objectives;
- Identifying potential treatment, resource recovery, and containment technologies that will satisfy these objectives;
- Screening the technologies based on their effectiveness, implementability, and cost; and
- Assembling technologies and their associated containment or disposal requirements into alternatives for the contaminated media at the site or for the different operable units or sectors of the site.

Alternatives can be developed to address contaminated medium, a specific area of the site, or the entire site. Once potential alternatives have been developed, it may be necessary to screen out certain options to reduce the number of alternatives that will be analyzed. The screening process involves evaluating alternatives with respect to their effectiveness, implementability, and cost. It is usually done on a general basis and with limited resources because the information necessary to fully evaluate the alternatives may not be complete at this point in the process.

2.1.2.4 *Treatability investigations*

Treatability investigations are conducted primarily to:

- Provide sufficient data to allow treatment alternatives to be fully developed and evaluated during the analysis phase and to support the remedial design (RD) of selected alternatives, and
- Reduce cost and performance uncertainties of treatment alternatives to acceptable levels so that a remedy can be selected.

2.1.2.5 *Detailed analysis*

Once sufficient data are available, alternatives are evaluated in detail with respect to the evaluation criteria that the regulatory agency recommends. The following criteria, used in a few developed countries, may be found useful in the evaluation, analysis, and selection of the technology:

- Overall protection of human health and the environment.
- Compliance with ARARs.

- Long-term effectiveness and permanence.
- Reduction of toxicity, mobility, or volume.
- Short-term effectiveness.
- Implementability.
- Cost.
- Acceptance by state and local authorities.
- Community acceptance.

The alternatives are analyzed individually against each criterion and then compared with one another to determine their respective strengths and weaknesses and to identify the key tradeoffs that must be balanced for the site. As mentioned earlier in this chapter, taking steps to prevent further migration of contaminants from the site deserves particular attention while selecting a remediation technology and during the remedial activities.

The results of the detailed analysis are summarized so that an appropriate remedy, consistent with the regulatory requirements, can be selected.

2.1.3 Record of treatment selection and decision

The record of treatment selected for the site or operable units of the site is created from information generated during the RI/FS. The record of treatment selection and decision (RTSD also known as ROD) also considers public comments and community concerns. For superfund sites, this is commonly referred to as the record of decision (ROD). The RTSD/ROD is a public document that explains which cleanup alternative, or alternatives, will be used to clean up a site.

2.1.4 Remedial design/remedial action

The RD/RA is based on the specifications described in the RTSD. RD is the phase of site cleanup where the technical specifications for cleanup remedies and technologies are designed. The detailed design specifications of the selected remedy are developed and reviewed by all of the stakeholders and the regulatory agency. Sometimes, the design specifications are changed and revised to meet the requirements of the regulatory agencies or the communities impacted by the waste site cleanup action. RA or the implementation phase involves the actual construction work in a site cleanup process following the RD phase.

Construction completion identifies successful completion of cleanup activities. A construction completion list could be developed to simplify the system of categorizing sites and to better communicate the successful completion of cleanup activities. Sites qualify to be in the construction completion list when:

- Any necessary physical construction is complete, whether or not final cleanup levels or other requirements have been achieved;
- The regulatory agency has determined that the response action should be limited to measures that do not involve further construction;
- The site qualifies for deletion from the listed hazardous waste sites; or
- Inclusion of a site on the construction completion list should not have any legal significance.

2.1.5 Considering climate change and health impacts of emissions during remedial actions

During the remedial technology applications, identification of specific pathways for migration of contaminants and appropriate controls of such pathways are emphasized in each of the technology applications in this book. The emissions from the contaminants from waste piles impact human health and the environment. In some cases, the reasons for high occurrences of certain diseases among residents living near waste sites were linked to their exposure to the contaminants due to migration of contaminants from industry-generated wastes that were kept in waste piles before any treatment. For example, 30 sites in Michigan and several sites in Indiana have reported contamination of drinking water with per- and polyfluoroalkyl substances (PFAS), which is a persistent chemical and has been linked to testicular and kidney cancer, liver damage, and developmental problems in children (Saenz, 2019).

The high levels of volatile organic contaminant (VOC) emissions from the waste piles during transportation, storage, and actual remediation activities also contribute toward the climate change by adding greenhouse gases to the atmosphere. As listed in Table 3.1, Chapter 3 of this book, several fluoro-methane and other halogenated and non-halogenated compounds are common VOCs found in the waste. These VOCs are mostly identified as greenhouse gas, and the fluoro-halogenated compounds are also considered ozone-depleting substances.

Although the factors affecting climate change and the related steps toward mitigating those factors are beyond the scope of this book, yet the serious concern about the climate change and the recent disasters due to the weather pattern changes and calamities all across the globe underscores the need to provide a synopsis for the readers to focus on this issue. A brief synopsis of the following findings by the European Commission (EC) and the National Aeronautics and Space Administration (NASA) should provide an understanding of the significance of this issue:

- Human activities are increasingly adding an enormous amount of greenhouse gases to those naturally occurring in the atmosphere, which is causing the greenhouse effect and global warming (EC, 2020).
- The current global average temperature is 0.85°C higher than it was in the late 19th century. Each of the past three decades has been warmer than any preceding decade since records began in 1850 (EC, 2020).
- Many of these gases occur naturally, but human activity is increasing the concentrations of some of them in the atmosphere, in particular (EC, 2020):
  - carbon dioxide (CO_2)
  - methane
  - nitrous oxide
  - fluorinated gases

  A report on climate change by NASA provided significant research data to reach almost identical conclusions as the European Commission (NASA, 2020).
- Figure 2.2 compares global surface temperature changes and the Sun's energy that Earth receives in watts per square meter since 1880. The lighter/thinner lines show the yearly levels while the heavier/thicker lines show the 11-year average trends. Eleven-year averages are used to reduce the year-to-year natural noise in the data, making the underlying trends more obvious.

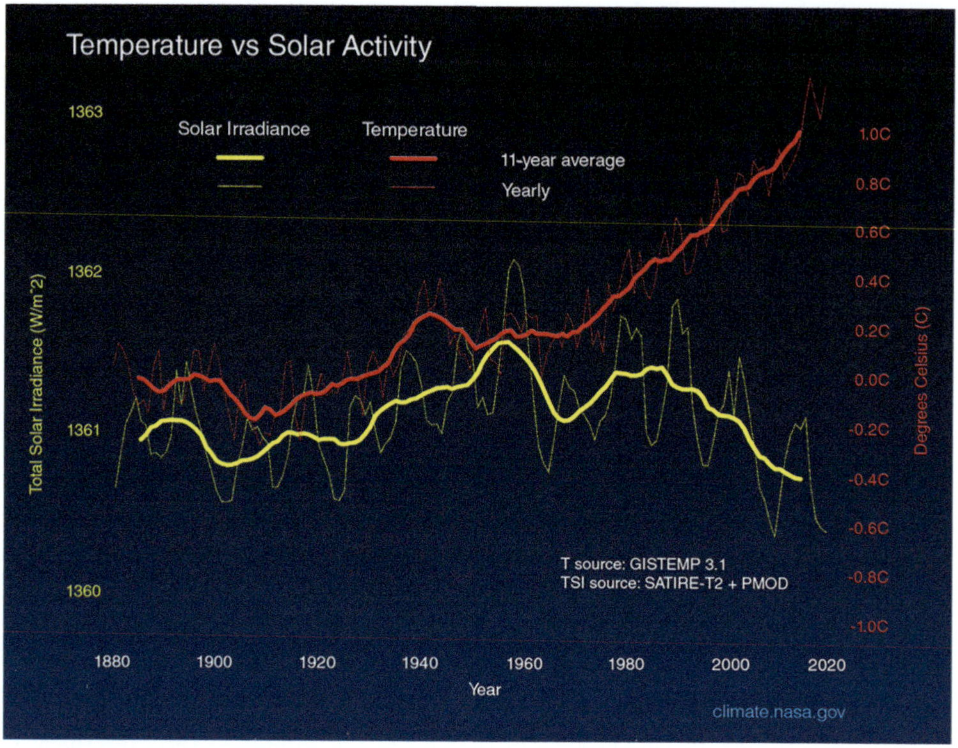

Figure 2.2 Global surface temperature changes and solar irradiance (NASA, 2020, Credit: NASA/JPL-Caltech.)

The amount of solar energy that Earth receives has followed the Sun's natural 11-year cycle of small ups and downs with no net increase since the 1950s. Over the same period, global temperature has risen markedly. It is therefore extremely unlikely that the Sun has caused the observed global temperature warming trend over the past half-century.

Although studies show that solar variability has played a role in past climate changes, several lines of evidence show that current global warming cannot be explained by changes in energy from the Sun, for example (NASA, 2020):

- Since 1750, the average amount of energy coming from the Sun either remained constant or increased slightly.
- It is evident from the data that greenhouse gases are trapping heat in the lower parts of the atmosphere causing the temperature rise. If the warming were caused by a more active Sun, then scientists would expect to see warmer temperatures in all layers of the atmosphere. Instead, they have observed a cooling in the upper atmosphere, and a warming at the surface and in the lower parts of the atmosphere.
- Climate models that include solar irradiance changes can't reproduce the observed temperature trend over the past century or more without including a rise in greenhouse gases.

Realizing the importance of controlling the greenhouse gas emissions, the focus of this book on cross-media transfer of contaminants and their control using the best management practices (BMPs)/options is emphasized. The value of proper control of cross-media transfer or migration/emission of contaminants during the implementation of various treatment technologies has twofold significance:

- Safeguarding the health of local residents and the people at or near the waste sites.
- Minimizing global warming and related disasters by using proper emission control at the waste sites during treatment, storage, and disposal activities.

In addition, recommendations on how to design and conduct remediation activities with minimal cross-media transfers of contaminants are provided in the treatment technology chapters of this book.

2.1.6 Operation and maintenance

In general, operation and maintenance (O&M) activities are conducted after the completion of the selected RA at a site. This ensures that the remedial operations in cleaning up the site are effective and operating properly. O&M activities protect the integrity of the selected remedy for a site. O&M measures are generally initiated by the state regulatory authorities after the RA objectives outlined in the RTSD have been met and are determined to be operational and functional based on the state and other regulatory agreements. RAs requiring O&M measures include landfill caps, gas collection systems, groundwater extraction treatment, groundwater monitoring, surface water treatment, and other similar actions that require long time period for the cleanup activity to attain the cleanup goal.

Once the O&M phase begins, the state or the site owner is responsible for maintaining the effectiveness of the remedy. O&M monitoring includes four components:

- Inspection.
- Sampling and analysis.
- Routine maintenance.
- Reporting.

O&M activities are usually required for sites where cleanup activities are proceeding through landfill/capping activities, groundwater treatment activities, or through monitoring of natural attenuation.

2.1.7 Closure and post-closure care

After the completion of remedial activities at a site the owners and operators of hazardous waste management units close these units in a manner that is protective of human health and the environment. These actions minimize any post-closure releases to the environment. The owners and operators are generally required by the regulatory agencies to submit closure plans to the agency for their hazardous waste management units. These plans are generally approved by the agency when they are determined to be compliant with the existing regulations.

The owners and operators of hazardous waste sites have to use various protective measures at their remediated sites during this post-closure period. For example, landfills are required to be covered with an impermeable cap designed to minimize infiltration of liquid into the unit; then owners or operators must conduct post-closure care (including maintenance of the cap and groundwater monitoring). Owners and operators of surface impoundments and waste piles must either remove or decontaminate all hazardous waste and constituents from the unit. Alternatively, they could leave the waste in place, install a final cover over the unit, and conduct post-closure care. Closure of land treatment facilities must be conducted in accordance with specific closure and post-closure care procedures designed to prevent leaching and migration of contaminants to surface or ground water. Owners and operators are generally required to remove some or all of the untreated waste from any type of unit at the time of closure. Owners and operators of fabricated units (e.g., tanks and containers) are required to remove or decontaminate all soils, structures, and equipment at closure. Owners and operators of tanks who are unable to do so are generally required to close the unit as a landfill and conduct post-closure care.

Where post-closure care is required, owners and operators are also required to comply with the requirements for installation of detection monitoring wells (e.g., one upgradient and three downgradient wells) The specific details of the system are worked out by the owners or operators in coordination with the regulatory agency in designing a monitoring program with site-specific indicator parameters. As a result, monitoring systems are specifically designed to monitor the constituents of concern at each individual site.

During the development of the post-closure requirements, generally the regulatory agencies ask the owner and operators of hazardous waste facilities to provide opportunities for public participation and community involvement. This approach is generally very effective and helps provide as much public participation as possible, besides generating invaluable feedback which is incorporated into the finalization of the post-closure plan.

2.2 COST AND ECONOMICS OF REMEDIATION

The cost and economics of remediation generally drive all remediation programs and activities in the US and other developed countries. This will probably be one of the most crucial factors to consider in any remediation activity in the developing countries. Due to significant variation in labor cost and local availability of the environmental treatment technologies covered in this book, no universal cost estimate for technologies could be furnished. However, some costs per unit volume on a few different treatment technologies, such as soil washing, and thermal treatment are provided in Chapters 5 and 6 in this book. In addition, where available, cost figures for some of the technologies are provided in this book in Chapter 11 – "Case studies of Treatment Technologies". Relative cost figures for various control practices are furnished in this book under Chapter 13 – "Monitoring and Control of Cross-Media Transfer of Contaminants During Cleanup Activities".

During the FS of a waste, the development of a site-specific cost and economics of remediation is emphasized for the developing countries. This should include various treatment alternatives, their cost, effectiveness, local availability, and suitability for the site when selecting or recommending the most appropriate remediation method.

REFERENCES

Dutta, S. (2002) *Environmental Treatment Technologies for Hazardous and Medical Wastes, Remedial Scope and Efficacy*. Tata McGraw Hill Publishing Company, New Delhi.

European Commission (EC). (2020) Causes of Climate Change. [Online] Available from: https://ec.europa.eu/clima/change/causes_en [Accessed 17th October, 2020].

NASA. (2020) Global Climate Change, Vital Signs of the Planet. [Online] Available from: https://climate.nasa.gov/causes/ [Accessed 17th October, 2020].

Saenz, E. (2019) EPA Creates Action Plan for Chemical Linked to Potentially Deadly Health Problems, Indiana Environmental Reporter, February. [Online] Available from: https://www.indianaenvironmentalreporter.org/posts/epa-creates-action-plan-for-chemical-linked-to-potentially-deadly-health-problems/ [Accessed 17th October, 2020].

USEPA. (1988) *A Guidance for Conducting Remedial Investigations and Feasibility Studies Under CERCLA*. Office of Emergency and Remedial Response, Washington, DC. EPA Directive No. 9355.3_01; NTIS PB89_184626, October 1988.

USEPA. (1991) *Guidance for Performing Preliminary Assessments Under CERCLA*. Office of Emergency and Remedial Response, Office of Solid Waste and Emergency Response, Washington, DC. EPA 9345.0_01A; NTIS PB92_963303.

USEPA. (1992) *The Hazard Ranking System Guidance Manual; Interim Final*. Office of Solid Waste and Emergency Response, Washington, DC. EPA 9345.1_07; NTIS PB92_963377, November 1992.

Chapter 3

Treatment technologies for municipal and industrial wastes with a focus on groundwater treatment

3.1 INTRODUCTION AND OVERVIEW

Remediation technologies encompass a wide range of treatment options that could be considered for cleanup of contaminated sites. As mentioned in Chapter 2, various factors, such as longterm effectiveness, permanence, reduction of toxicity, mobility, or volume, short-term effectiveness, implementability, cost, acceptance by State and local authorities, and community acceptance are evaluated and analyzed to determine the suitability of a particular technology at a specific site. In general, remediation treatment includes technologies that are used to treat sites, which have either contaminated soil, or debris/solids, or sediments, or groundwater or any combination of these. Since the most commonly encountered problems in waste sites involve solids, debris, sludges or sediments needing treatment or disposal, technologies used for solid wastes will be primarily covered in details in this book. Some of these treatment technologies are also used for groundwater, either exclusively or in combination with other methodologies. Such applicability of these technologies will be identified and addressed under those specific technologies.

This chapter categorizes remedial treatment options for various types of municipal or industrial wastes containing soil, sediments, or sludges. It provides a comprehensive list of volatile organic compounds (VOCs), semi-volatile organic compounds (SVOCs), metals, polynuclear aromatic hydrocarbons (PAHs), radionuclides, pesticides, and other inorganic contaminants. The term polycyclic aromatic hydrocarbon (PAH) is commonlyused since most of the PAHs have the presence of two or more cyclic structures fused toeach other whereas the term 'polynuclear' describes broadly the presence of more than one atom. Most of the municipal and industrial waste site remediation involves solids, sediments, and sludges requiring a spectrum of treatment technologies that are suitable for some specific site conditions. It will be a colossal task to cover the details on each of those technologies within the bounds of this book. To simplify that task, the treatment technologies for soil/solids, sediments, and sludges have been grouped based on common features and similarities in treating contaminants.

The treatment technologies for soil/solids, sediments, and sludges are grouped into seven technology categories and addressed in detail covering each group separately in one chapter beginning from Chapters 4 to 10 of this book as mentioned later in Section 3.2.

The following subsection provides an overview of the groundwater treatment technologies.

3.2 GROUNDWATER TREATMENT TECHNOLOGIES

Most of the groundwater cleanup systems involve extraction of contaminated groundwater, above ground treatment and removal of contaminants, and surface discharge or reinjection

DOI: 10.1201/9781003004066-3

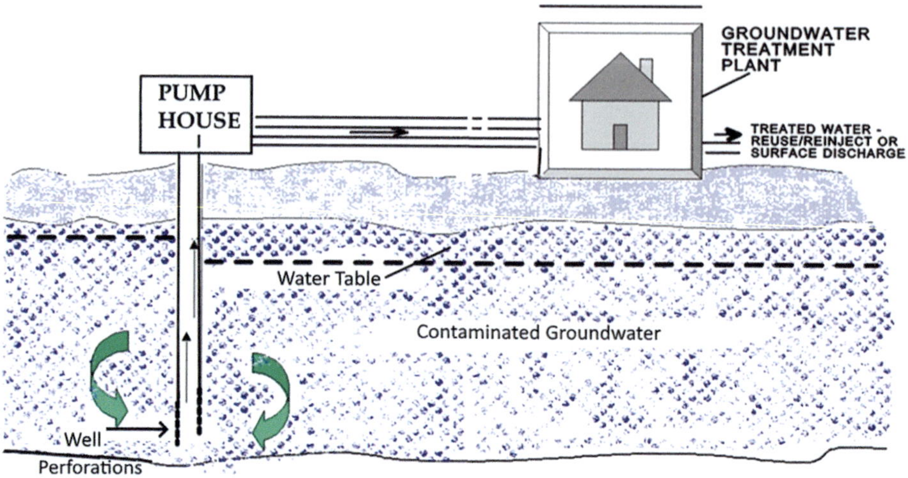

Figure 3.1 A schematic groundwater treatment system (Dutta, 2002).

of treated water. Figure 3.1 shows a schematic groundwater treatment system involving pumping/extraction and treatment of contaminated groundwater. In some cases, in situ treatment or treatment in-place is also conducted where appropriate treatment ingredients are injected into the contaminated groundwater. Sometimes, natural attenuation is also used as a remedy where the natural degradation rate of the contaminants is high, and it is considered the most safe and feasible option for the site. Further discussion of various types of groundwater treatment is limited by the scope of this book. However, readers may look into various publications cited under reference at the end of this chapter for further details on in situ treatment or natural attenuation of contaminated groundwater. It should be noted that treatment technologies that are applicable for groundwater could also be applied for *Surface water* or *leachate* with similar contaminant characteristics. For details on surface water or leachate treatments, readers may find USEPA 2020 listed under reference as a useful guide.

Groundwater often gets contaminated with water-soluble substances found in overlying soils. Many of the parameters needed to assess different treatment options for the contaminated groundwater are similar, which include pH, total organic carbon (TOC), biochemical oxygen demand (BOD), chemical oxygen demand (COD), conductivity, oil and grease content, and any other chemical contaminants of concern. Identification and quantification of the major contaminants in soil layers above the aquifer and a complete aquifer characterization are also considered as major requirements for evaluating the best treatment option for the contaminated groundwater. Additional water quality monitoring data elements include hardness, ammonia, total dissolved solids, and metals content (e.g., iron and manganese). Knowledge of the site conditions and history may contribute to selecting a list of contaminants and cost-effective analytical methods.

Technologies that are commonly used to treat groundwater can be grouped into three main categories depending upon the type of contaminants that are effectively treated by them. These three categories are as follows:

- VOCs
- SVOCs
- Inorganic (metals) contaminants.

A list of commonly encountered VOCs and SVOCs and inorganic contaminants is provided in Table 3.1.

3.2.1 Properties and behavior of VOCs

While evaluating the remedy for a chemical compound, it is important to consider whether the compound is halogenated or nonhalogenated. A halogenated compound is one onto which a halogen (e.g., fluorine, chlorine, bromine, or iodine) has been attached. Typical halogenated and nonhalogenated VOCs, SVOCs, metals, and other contaminants are listed in Table 3.1. The nature of the halogen bond and the halogen itself can significantly affect the performance of a technology or require more extensive treatment than for nonhalogenated compounds.

As an example, let us consider bioremediation. In general, halogenated compounds are less amenable to this form of treatment than nonhalogenated compounds. Moreover, the more halogenated the compound (i.e., the more halogens attached to it), the more recalcitrant it is toward biodegradation. As another example, incineration of halogenated compounds requires specific off-gas and scrubber water treatment for the halogen in addition to the normal controls that are implemented for nonhalogenated compounds (USEPA, 1993).

Subsurface contamination by VOCs potentially exists in four phases:

- Gaseous phase: Contaminants present as vapors in unsaturated zones.
- Solid phase: Contaminants in liquid form adsorbed on soil particles in both saturated and unsaturated zones.
- Aqueous phase: Contaminants dissolved into pore water according to their solubility in both saturated and unsaturated zones.
- Immiscible phase: Contaminants present as non-aqueous phase liquids (NAPLs) primarily in unsaturated zone.

One or more of the fluid phases (gaseous, liquid, aqueous, or immiscible) may occupy the pore spaces in the unsaturated zone. Residual bulk liquid may be retained by capillary attraction in the porous media (i.e., NAPLs are no longer a continuous phase but are present as isolated residual globules).

Residual saturation of bulk liquid may occur through a number of mechanisms. Volatilization from residual saturation or bulk liquid into the unsaturated pore spaces produces a vapor plume. Lateral migration of this vapor plume is independent of groundwater movement and may occur as a result of both advection and diffusion. Advection is the process by which the vapor plume contaminants are transported by the movement of air and may result from gas pressure or gas density gradients. Diffusion is the movement of contaminants from areas

Table 3.1 List of typical VOCs, SVOCs, metals, and other contaminants (Dutta, 2002)

Volatile organic compounds (VOCs)/metals/radionuclides/inorganics	*Semivolatile organic compounds (SVOCs) /PAHs/pesticides*
Halogenated VOCs:	*Halogenated SVOCs:*
Bromodichloromethane	Bis(2-chloroethoxy)ether
Bromoform	1,2-Bis(2-chloroethoxy) ethane
Bromomethane	Bis(2-chloroethoxy) methane
Carbon tetrachloride	Bis(2-chloroethoxy) phthalate
Chlorodibromomethane	Bis(2-chloroethyl)ether
Chloroethane	Bis(2-chloroisopropyl) ether
Chloroform	4-Bromophenyl phenyl ether
Chloromethane	4-Chloroaniline
Chloropropane	p-Chloro-m-cresol
Cis-1,2-dichloroethylene	2-Chloronaphthalene
Cis-1,3-dichloropropene	1,3-Dichlorobenzene
Dibromomethane	1,4-Dichlorobenzene
1,1-Dichloroethylene	3,3-Dichlorobenzidine
Dichloromethane	2,4-Dichlorophenol
1,2-Dichloropropane	Hexachlorobenzene
Ethylene dibromide	Hexachlorobutadiene
Fluorotrichloromethane (Freon 11)	Hexachlorocyclopentadiene
Hexachloroethane	Pentachlorophenol (PCP)
Methylene chloride	Polychlorinated biphenyls (PCBs)
Monochlorobenzene	Tetrachlorophenol
1,1,2,2-Tetrachloroethane	1,2,4-Trichlorobenzene
Tetrachloroethylene	2-Chlorophenol
Perchloroethylene (PCE)	4-Chlorophenyl phenylether
1,2-Trans-dichloroethylene	1,2-Dichlorobenzene
1,1-Dichloroethane	2,4,5-Trichlorophenol
1,2-Dichloroethane	2,4,6-Trichlorophenol
1,2-Dichloroethene	
Trichloroethylene (TCE)	
1,2,2-trifluoroethane (Freon 113)	
Trans-1,3-dichloropropene	
1,1,1-Trichloroethane	
1,1,2-Trichloroethane	
Vinyl chloride	
Nonhalogenated VOCs:	*Nonhalogenated SVOCs*
Acetone	Benzidine
Acrolein	Benzoic Acid
Acrylonitrile	Benzyl alcohol
n-Butyl alcohol	Bis(2-ethylhexyl)phthalate
Carbon disulfide	Butyl benzyl phthalate
Cyclohexanone	Dibenzofuran
Ethyl acetate	Di-n-butyl phthalate
Ethyl ether	Di-n-octyl phthalate
Isobutanol	Diethyl phthalate
Methanol	Dimethyl phthalate
Methyl ethyl ketone (MEK)	4,6-Dinitro-2-methylphenol

(Continued)

Table 3.1 (Continued) List of typical VOCs, SVOCs, metals, and other contaminants

Volatile organic compounds (VOCs)/metals/radionuclides/inorganics	*Semivolatile organic compounds (SVOCs) /PAHs/pesticides*
Methyl isobutyl ketone 4-Methyl-2-pentanone Styrene Tetrahydrofuran Vinyl acetate *Metals* Aluminum, Antimony, Arsenic Aluminum, Antimony, Arsenic Barium, Beryllium, Bismuth, Boron Cadmium, Calcium, Chromium, Cobalt Copper, Iron, Lead Magnesium Manganese Mercury Metallic cyanides Nickel Potassium Selenium Silver Sodium Thallium Tin Titanium Vanadium Zinc *Radionuclides* Americium-241 Cesium-134, -137 Cobalt-60 Europium-152, -154, -155 Plutonium-238, -239 Radium-224, -226 Strontium-90 Technetium-99 Thorium-228, -230, -232 Uranium-234, -235, -238 *Other inorganic contaminants* Asbestos (non-volatile) Cyanide Fluorine	2,4-Dinitrophenol 1,2-Diphenylhydrazine Isophorone 2-Nitroaniline 3-Nitroaniline 4-Nitroaniline 2-Nitrophenol 4-Nitrophenol n-Nitrosodimethylamine n-Nitrosodiphenylamine n-Nitrosodi-n-propylamine Phenyl naphthalene *Polynuclear Aromatic Hydrocarbons (PAHs)* Acenaphthene Acenaphthylene Anthracene Benzo(a)anthracene Benzo(a)pyrene Benzo(b)fluoranthene Benzo(k)fluoranthene Chrysene Fluoranthene Fluorene Indeno(1,2,3-cd) pyrene 2-Methylnaphthalene Naphthalene Phenanthrene, Pyrene *Pesticides* Aldrin BHC-alpha BHC-beta BHC-delta BHC-gamma Chlordane 4,4′-DDD 4,4′-DDE 4,4′-DDT Dieldrin, Endosulfan I, Endosulfan II Endosulfan sulfate, Endrin Endrin aldehyde Ethion, Ethyl parathion Heptachlor, Heptachlor epoxide Malathion, Methylparathion Parathion Toxaphene

of high vapor concentrations to areas of lower vapor concentrations. Volatilization from contaminated groundwater also may produce a vapor plume of compounds with high vapor pressures and high aqueous solubilities.

Dissolution of contaminants from residual saturation or bulk liquid into water may occur either in the unsaturated or in the saturated portions of the subsurface with the contamination then moving with the water. Even low-solubility organics may be present at low concentrations dissolved in water.

Insoluble organic contaminants may be present as NAPLs. Dense NAPLs (DNAPLs) have a specific gravity greater than 1 and will tend to sink to the bottom of surface waters and groundwater aquifers. Light NAPLs (LNAPLs) will float on top of surface water and groundwater. In addition, DNAPLs and LNAPLs may adhere to the soil through the capillary fringe and may be found on top of water in temporary or perched aquifers in the vadose zone.

3.2.1.1 Treatment technologies for VOCs in groundwater

The most commonly used technologies to treat VOCs in groundwater are air stripping and liquid-phase carbon adsorption. These are both ex situ technologies requiring groundwater extraction.

Air stripping involves the transfer of volatile contaminants from water to air. This process is typically conducted in a packed tower or in an aeration tank. The generic packed tower air stripper includes a spray nozzle at the top of the tower to distribute contaminated water over the packing in the column, a fan to force air countercurrent to the water flow, and a sump at the bottom of the tower to collect decontaminated water. Auxiliary equipment that can be added to the basic air stripper includes a feed water heater (normally not included because of the high cost) and an air heater to improve removal efficiencies, automated control systems with sump level switches and safety features such as differential pressure monitors, high sump-level switches and explosion proof components, and discharge air treatment systems such as activated carbon units, catalytic oxidizers, or thermal oxidizers. Packed tower air strippers are installed either as permanent installations on concrete pads or as temporary installations on skids or on trailers.

In-well air stripping systems create a circulation pattern in the aquifer by drawing water into and pumping it through the wells and then reintroducing the water into the aquifer without bringing it above ground. The well is double-cased with hydraulically separated upper and lower screened intervals within the aquifer. The system can be configured with an upward in-well flow or a downward in-well flow. The most common configurations involve the injection of air into the inner casing, decreasing the density of the groundwater and allowing it to rise. Through this system, volatile contaminants in the groundwater are transferred from the dissolved phase to the vapor phase by the rising air bubbles. Contaminated vapors can be drawn off and treated above ground or discharged into the vadose zone (USEPA, 2020).

Carbon adsorption (aqueous) is a technology in which groundwater is pumped through a series of vessels containing activated carbon to which dissolved contaminants adsorb. This type of direct adsorption of dissolved contaminants from the liquid is generally referred to as the liquid-phase carbon adsorption system. When the concentration of contaminants in the effluent from the bed exceeds a certain level, the carbon can be regenerated in place; removed and regenerated at an off-site facility; or removed and disposed of. Carbon used for explosives- or metals-contaminated groundwater must be removed and properly disposed of. Adsorption by activated carbon has a long history of use in treating municipal, industrial, and hazardous wastes.

Carbon adsorption is also commonly used to remove the off-gas or vapors from groundwater treatment systems. This is also very similar to the liquid-phase system in which pollutants are removed from air by physical adsorption onto the carbon grain. Carbon is "activated" for this purpose by processing the carbon to create porous particles with a large internal surface area (300–2,500 m^2/gm of carbon) that attracts and adsorbs organic molecules as well as certain metal and other inorganic molecules.

Commercial grades of activated carbon are available for specific use in vapor-phase applications. The granular form of activated carbon is typically used in packed beds through which the contaminated air flows until the concentration of contaminants in the effluent from the carbon bed exceeds an acceptable level. Granular activated carbon systems typically consist of one or more vessels filled with carbon connected in series and/or parallel operating under atmospheric, negative, or positive pressure. The most commonly used technologies to treat VOCs in groundwater are air stripping and liquid-phase carbon adsorption. These are both ex situ technologies requiring groundwater extraction.

The carbon can then be regenerated in place, regenerated at an off-site regeneration facility, or disposed of, depending on economic considerations. One such example is Lockheed Martin's Groundwater treatment plant in Burbank, California where moderate vacuum steam stripping system is used to remove trichloroethylene (TCE) from groundwater, which is then adsorbed by the carbon beds. A photograph of the carbon beds at this site is shown in Figure 3.2.

There are three main technologies that are most commonly used to treat VOCs in air emissions/off-gases. These are carbon adsorption, catalytic oxidation, and thermal oxidation. A brief description of each of these technologies are provided below:

Carbon adsorption (Gaseous) is also a remediation technology in which pollutants are removed from air by physical adsorption onto the carbon grain. Carbon is "activated" for

Figure 3.2 Carbon adsorption system at Lockheed Martin's site in Burbank, CA (Dutta, 2002).

this purpose by processing the carbon to create porous particles with a large internal surface area that attracts and adsorbs organic molecules as well as certain metal and other inorganic molecules.

Catalytic oxidation is a relatively new alternative for the treatment of VOCs in air streams resulting from remedial operations. VOCs are thermally destroyed at temperatures typically ranging from 600°F to 1,000°F by using a solid catalyst. First, the contaminated air is directly preheated (electrically or, more frequently, using natural gas or propane) to reach a temperature necessary to initiate the catalytic oxidation of the VOCs. Then the preheated VOC-laden air is passed through a bed of solid catalysts where the VOCs are rapidly oxidized.

In most cases, the process can be enhanced to reduce auxiliary fuel costs by using an air-to-air heat exchanger to transfer heat from the exhaust gases to the incoming contaminated air. Typically, about 50% of the heat of the exhaust gases is recovered. Based on VOC concentrations, the recovered heat may be sufficient to sustain oxidation without additional fuel. Catalyst systems that are used to oxidize VOCs typically use metal oxides such as nickel oxide, copper oxide, manganese dioxide, or chromium oxide. Noble metals such as platinum and palladium may also be used. However, in the majority of remedial applications, nonprecious metals (e.g., nickel, copper, or chromium) are used. Most commercially available catalysts are proprietary.

Thermal oxidation equipment is used for destroying contaminants in the exhaust gas from air strippers and SVE systems. Probably, fewer than 100 oxidizers have been sold to treat air stripper effluents; most of these units are rated less than 600 std. cubic feet/min (SCFM) or 17 std. cubic m/min. Typically, the blower for the air stripper or the vacuum extraction system provides sufficient positive pressure and flow for thermal oxidizer operation.

Thermal oxidation units are typically single chamber, refractory-lined oxidizers equipped with a propane or natural gas burner and a stack. Lightweight ceramic blanket refractory is used because many of these units are mounted on skids or trailers. Thermal oxidizers are often equipped with heat exchangers where combustion gas is used to preheat the incoming contaminated gas. If gasoline is the contaminant, heat exchanger efficiencies are limited to 25 to 35% and preheat temperatures are maintained below 530°F to minimize the possibility of ignition occurring in the heat exchanger. Flame arrestors are always installed between the vapor source and the thermal oxidizer. Burner capacities in the combustion chamber range from 0.5 to 2 million BTUs per hour (0.15 W to 586.17 Kw). Operating temperatures range from 1,400°F to 1,600°F, and gas residence times are typically 1 second or less.

3.2.2 Properties and behavior of SVOCs

SVOCs may be found in sites which have been or are typically used for burn pits, chemical manufacturing plants and disposal areas, contaminated marine sediments, disposal wells and leach fields, electroplating/metal finishing shops, firefighting training areas, hangars/aircraft maintenance areas, landfills and burial pits, leaking collection and system sanitary lines, leaking storage tanks, radiologic/mixed waste disposal areas, oxidation ponds/lagoons, pesticide/herbicide mixing areas, solvent degreasing areas, surface impoundments, and vehicle maintenance areas and wood-preserving sites.

An important consideration when evaluating a remedy for a SVOC compound is to check whether the contaminant is halogenated or nonhalogenated. A halogenated compound is one onto which a halogen (e.g., fluorine, chlorine, bromine, or iodine) has been attached. Typical halogenated and nonhalogenated SVOCs are listed in Table 3.1. The nature of the halogen bond and the halogen itself can significantly affect the performance of a technology or require more extensive treatment than for nonhalogenated compounds.

As an example, consider bioremediation. In general, halogenated compounds are less amenable to this form of treatment than nonhalogenated compounds. In addition, the more halogenated the compound (i.e., the more halogens attached to it), the more defiant it is toward biodegradation. As another example, incineration of halogenated compounds requires specific off-gas and scrubber water treatment for the halogen in addition to the normal controls that are implemented for nonhalogenated compounds.

Therefore, the vendor of the technology being evaluated must be informed whether the compounds to be treated are halogenated or nonhalogenated. In most instances, the vendor needs to know the specific compounds involved so that modifications to technology designs can be made, where appropriate, to make the technology successful in treating halogenated compounds.

Subsurface contamination by SVOCs potentially exists in four phases:

- Gaseous phase: contaminants present as vapors in the saturated zones.
- Solid phase: contaminants adsorbed or partitioned onto the soil or aquifer material in both saturated and unsaturated zones.
- Aqueous phase: contaminants dissolved into pore water according to their solubility in both saturated and unsaturated zones.
- Immiscible phase: contaminants present as NAPLs primarily in the saturated zones.

One or more of the three fluid phases (gaseous, aqueous, or immiscible) may occupy the pore spaces in the unsaturated zone. Residual bulk liquid may be retained by capillary attraction in the porous media (i.e., NAPLs are no longer a continuous phase but are present as isolated residual globules).

Contaminant flow may occur through a number of mechanisms. Volatilization from residual saturation or bulk liquid into the unsaturated pore spaces produces a vapor plume. While the degree of volatilization from SVOCs is much less than for VOCs, this process still occurs.

Similar to VOCs, dissolution of SVOCs from residual saturation or bulk liquid into water may occur either in the unsaturated or in the saturated portions of the subsurface with the contamination then moving with the water. Even low-solubility organics may be present at low concentrations dissolved in water.

Insoluble or low solubility organic contaminants may be present as NAPLs. Dense non-aqueous phase liquids (DNAPLs) will tend to sink to the bottom of surface waters and groundwater aquifers. Light non-aqueous phase liquids (LNAPLs) will float on top of surface water and groundwater. In addition, LNAPLs may adhere to the soil through the capillary fringe and may be found on top of water in temporary or perched aquifers in the vadose zone.

Properties and behavior of specific SVOC contaminants and contaminant groups are discussed below:

- *PAHs*: PAHs are generally biodegradable in soil systems. Lower molecular weight PAHs are transformed much more quickly than higher molecular weight PAHs. The less degradable, higher molecular weight compounds have been classified as carcinogenic PAHs (cPAHs). Benzo[a]pyrene (BaP) and benzo[b]fluoranthene are the most carcinogenic, followed by benzo[j]fluoranthene. Therefore, the least degradable fraction of PAH contaminants in soils is generally subject to the most stringent cleanup standards. This presents some difficulty in achieving cleanup goals with bioremediation systems.

Lower molecular weight PAH components are more water-soluble than higher molecular weight PAHs. Readily mobilized compounds, such as naphthalene, phenanthrene, and anthracene, are slightly water-soluble. Persistent PAHs, such as chrysene and benzo(a)pyrene, present even lower water solubilities. Pyrene and fluoranthene are exceptions because these compounds are more soluble than anthracene, but are not appreciably metabolized by soil microorganisms. Other factors affect PAH persistence such as insufficient bacterial membrane permeability, lack of enzyme specificity, and insufficient aerobic conditions. PAHs may undergo significant interactions with soil organic matter.

Intermediate PAH degradation products (metabolites) in soil treatment systems may also display toxicity. Complete mineralization of PAHs is slow; intermediates may remain for substantial periods of time.

- *PCBs*: PCBs encompass a class of chlorinated compounds that includes up to 209 variations or congeners with different physical and chemical characteristics. PCBs were commonly used as mixtures called aroclors. The most common aroclors are Aroclor-1254, Aroclor-1260, and Aroclor-1242. PCBs alone are not usually very mobile in subsurface soils or water; however, they are typically found in oils associated with electrical transformers or gas pipelines or sorbed to soil particles, which may transport the PCBs by wind or water erosion.
- *Pentachlorophenol (PCP)*: PCP is a contaminant found at many wood-preserving sites. PCP does not decompose when heated to its boiling point for extended periods of time. Pure PCP is chemically rather inert. The chlorinated ring structure tends to increase stability, but the polar hydroxyl group facilitates biological degradation. All monovalent alkali metal salts of PCP are very soluble in water. The protonated (phenolic) form is less soluble, but this degree of solubility is still significant from an environmental standpoint. PCP can also volatilize from soils. It is denser than water, but the commonly used solution contains PCP and petroleum solvents in a mixture less dense than water. Therefore, technical grade PCP floats on the top of groundwater as a LNAPL.
- *Pesticides:* The term "pesticide" is applied to literally thousands of different, specific chemical-end products. Pesticides include insecticides, fungicides, herbicides, acaricides, nematodicides, and rodenticides. There are several commonly used classification criteria that can be used to group pesticides for the purposes of discussion. Conventional methods of classifying pesticides base categorization on the applicability of a substance or product to the type of pest control desired. (For example, DDT is used typically as an insecticide.) The RCRA hazardous waste classification system is based on waste characterization and sources. Neither of these classification formats has any bearing on applicable treatment technologies for remediating pesticide-contaminated sites, and hence it is not specifically addressed in this book.

3.2.2.1 Common treatment technologies for SVOCs in groundwater

In addition to the general data requirements, it may be necessary to know other subsurface information to remediate semi-volatile organics in water. Treatability studies may be required to determine the contaminant biodegradability for any biodegradation technologies. Treatability studies are also necessary to ensure that the contaminated groundwater can be treated effectively at the design flow. A subsurface geologic characterization would be particularly useful for any isolation or stabilization technologies. Groundwater models are also often needed to predict flow characteristics, changes in contaminant mixes and concentrations, capture zones and times to reach cleanup levels.

The most commonly used ex situ treatment technologies for SVOCs in groundwater and surface water include carbon adsorption and UV oxidation. In situ treatment technologies are not widely used. Sometimes, the groundwater and surface water concentrations are not sufficiently high to support biological processes; however, for leachate treatment, biological process could be applied.

Carbon adsorption, as described before, for VOCs is also considered to be a commonly used technology to treat SVOCs in groundwater. In this system, the groundwater is pumped through a series of vessels containing activated carbon to which dissolved contaminants are adsorbed. When the concentration of contaminants in the effluent from the bed exceeds a certain level, the carbon can be regenerated in place; removed and regenerated at an off-site facility; or removed and disposed of. Carbon used for explosives- or metals-contaminated groundwater must be removed and properly disposed of. Adsorption by activated carbon has a long history of use in treating municipal, industrial, and hazardous wastes.

Ultra violet (UV) oxidation is a destruction process that oxidizes organic and explosive constituents in wastewaters by the addition of strong oxidizers and irradiation with intense UV light. The oxidation reactions are catalyzed by UV light, while ozone (O_3) and/or hydrogen peroxide (H_2O_2) are commonly used as oxidizing agents. The final products of oxidation are carbon dioxide, water, and salts. The main advantage of UV oxidation is that organic contaminants can be converted to relatively harmless carbon dioxide and water by hydroxyl radicals generated during the process. UV oxidation processes can be configured in batch or in continuous flow modes. Catalyst addition may enhance the performance of the system.

3.2.3 Properties and behavior of fuels

Fuel is another very common source of site contamination. Sites where fuel contaminants may be found include aircraft areas, burn pits, chemical disposal areas, contaminated marine sediments, disposal wells and leach fields, firefighting training areas, hangars/aircraft maintenance areas, landfills and burial pits, leaking storage tanks, solvent degreasing areas, surface impoundments, and vehicle maintenance areas. Typical fuel contaminants encountered at many sites include the following compounds as shown in the list below:

A BULLET LIST OF TYPICAL FUEL CONTAMINANTS

- Anthracene
- Benz(a)anthracene
- Benzene
- Benzo(b)fluoranthene
- Benzo(k)fluoranthene
- Benzo(g, h, i)perylene
- Methylcyclopentane
- 2-Methylheptane
- 3-Methylheptane
- 3-Methylhexane
- Methylnaphthalene
- 2-Methylnaphthalene
- 1,2,4,5-Tetramethylbenzene
- Toluene
- 1,2,4-Trimethylbenzene
- 1,3,5-Trimethylbenzene
- 1,2,4-Trimethyl-5-ethylbenzene
- 2,2,4-Trimethylheptane

(Continued)

- Benzo(a)pyrene
- Chrysene
- Cis-2-butene
- Creosols
- Cyclohexane
- Cyclopentane
- Dibenzo(a, h)anthracene
- 2,3-Dimethylbutane
- 3,3-Dimethyl-1-butene
- Dimethylethylbenzene
- 2,2-Dimethylheptane
- 2,2-Dimethylhexane
- 2,2-Dimethylpentane
- 2,3-Dimethylpentane
- 2,4-Dimethylphenol
- Ethylbenzene
- 3-Ethylpentane
- Fluoranthene
- Fluorene
- Ideno(1,2,3-c, d)pyrene
- Isobutane
- Isopentane
- 2-Methyl-1,3-butadiene
- 3-Methyl-1,2-butadiene
- 2-Methyl-butene
- 2-Methyl-2-butene
- 3-Methyl-1-butene
- 2-Methylpentane
- 3-Methylpentane
- 3-Methyl-1-pentene
- 2-Methylphenol
- 4-Methylphenol
- Methylpropylbenzene
- m-Xylene
- Naphthalene
- n-Butane
- n-Decane
- n-Dodecane
- n-Heptane
- n-Hexane
- n-Hexylbenzene
- n-Nonane
- n-Octane
- n-Pentane
- n-Propylbenzene
- n-Undecane
- o-Xylene
- 1-Pentene
- Phenanthrene
- Phenol
- Propane
- p-Xylene
- Pyrene
- Pyridine
- 2,3,4-Trimethylheptane
- 3,3,5-Trimethylheptane
- 2,4,4-Trimethylhexane
- 2,3,4-Trimethylhexane
- 2,2,4-Trimethylpentane
- 2,3,4-Trimethylpentane
- Trans-2-butene
- Trans-2-pentene

Information presented for VOCs and SVOCs may also be appropriate for many of the fuel contaminants presented in this subsection. As previously discussed for VOCs and SVOCs, an important consideration when evaluating a remedy is whether the compound is halogenated or nonhalogenated. Fuel contaminants are generally nonhalogenated. A halogenated compound is one onto which a halogen (e.g., fluorine, chlorine, bromine, or iodine) has been attached. The nature of the halogen bond and the halogen itself can significantly affect the performance of a technology or require more extensive treatment than for nonhalogenated compounds.

Contamination by fuel contaminants in the unsaturated zone exists in four phases: vapor in the pore spaces; sorbed to subsurface solids; dissolved in water; or as NAPL. The nature and extent of transport are determined by the interactions among contaminant transport properties (e.g., density, vapor pressure, viscosity, and hydrophobicity) and the subsurface environment (e.g., geology, aquifer mineralogy, and groundwater hydrology). Most fuel-derived contaminants are less dense than water and can be detected as floating pools (LNAPLs) on the water table.

Typically, after a spill occurs, LNAPLs migrate vertically in the subsurface until residual saturation depletes the liquid or until the capillary fringe above the water table is reached. Some spreading of the bulk liquid occurs until pressure from the infiltrating liquid develops sufficiently to penetrate to the water table. The pressure of the infiltrating liquid pushes the spill below the surface of the water table. Bulk liquids that are less dense than water spread laterally and float on the surface of the water table, forming a mound that becomes compressed into a spreading lens.

As the plume of dissolved constituents moves away from the floating bulk liquid, interactions with the soil particles affect dissolved concentrations. Compounds that are more attracted to the aquifer material move at a slower pace than the groundwater and are found closer to the source; compounds that are less attracted to the soil particles move most rapidly and are found at the leading edge of a contaminant plume.

LNAPL compounds that are more volatile readily partition into the air phase. A soil gas sample collected from an area contaminated by vapor-phase transport typically contains relatively higher concentrations of the more volatile compounds than one contaminated by groundwater transport. Vapor-phase transport can be followed by subsequent dissolution in groundwater. Alternatively, aqueous-phase contaminants with high Henry's law constants are likely to volatilize into the pore spaces.

As a result of gas density gradients, density-driven flow of the vapor plume may occur for compounds with vapor densities greater than air. Toluene, ethylbenzene, xylenes, and naphthalene are less dense than water and unlikely to move by density-driven flow. However, they may be capable of diffusive transport, causing vapor plumes to move away from residual saturation in the unsaturated zone. Residual saturation is the portion of the liquid contaminant that remains in the pore spaces as a result of capillary action after the NAPL moves through the soil. Volatilization from contaminated groundwater also may produce a vapor plume of compounds with high vapor pressures and high aqueous solubilities. Dissolution of fuel contaminants from residual saturation or bulk liquid into water may occur either in the unsaturated or in the saturated portions of the subsurface. The contamination thus moves with the water. Because the solubility of fuels is relatively low, contaminant dissolution from NAPL under laminar flow conditions typical of aquifers is mass-transfer limited, requiring decades for dissolution and producing a dilute waste-stream of massive volume.

Some of the groundwater treatment technologies, such as UV oxidation and air stripping, may be suitably used to treat liquid wastes from hospitals containing highly contagious waste droplets including nano-scale virus (e.g., COVID-19) particles.

3.2.3.1 Treatment technologies for fuels in groundwater

In addition to the general data requirements, it may be necessary to know other subsurface information to remediate fuels in groundwater. Any biodegradation technology may require treatability testing to characterize contaminant biodegradability and nutrient content. A subsurface geologic characterization would be particularly important to characterize the migration of NAPLs. Recovery tests are usually necessary to design a product/groundwater pumping scheme that will ensure that the nonaqueous fuel layer can be recovered and that contaminated groundwater can be treated effectively at the design flow. Groundwater models are also often needed to predict flow characteristics, changes in contaminant mixes and concentrations, capture zones, and times to reach cleanup levels.

Technologies most commonly used to treat fuels in groundwater include air stripping, carbon adsorption, and oil–water separator and recovery. These are all ex situ treatment technologies requiring groundwater extraction.

Air stripping is commonly used for treating fuel-contaminated groundwater. In addition to the generic air stripping treatment of groundwater, some specific features of air stripping treatment for effectively removing fuel contaminants from groundwater are provided here. Standard air stripping process is typically conducted in a packed tower or in an aeration tank. The generic packed tower air stripper includes a spray nozzle at the top of the tower to distribute contaminated water over the packing in the column, a fan to force air countercurrent to the water flow, and a sump at the bottom of the tower to collect decontaminated water. Packed tower air strippers are installed either as permanent installations on concrete pads, on a skid, or on a trailer. Auxiliary equipment can be added to the basic air stripper to effectively remove fuel contaminants as vapors. These additional features can include automated control systems with sump level switches and safety features such as differential pressure monitors, high sump level switches and explosion proof components, and discharge air treatment systems such as activated carbon units, catalytic oxidizers, or thermal oxidizers.

In-Well Air Stripping systems can also be used for fuel contaminated groundwater where the volatile contaminants in the ground water are transferred from the dissolved phase to the vapor phase by the rising air bubbles during the in-well air stripping operation described earlier in this chapter. Contaminated vapors can be drawn off and treated above ground or discharged into the vadose zone (USEPA, 2020).

Liquid phase carbon adsorption is a full-scale technology in which groundwater is pumped through a series of vessels containing activated carbon to which dissolved contaminants are adsorbed. When the concentration of contaminants in the effluent from the bed exceeds a certain level, the carbon can be regenerated in place; removed and regenerated at an off-site facility; or removed and disposed of. Adsorption by activated carbon has a long history of use in treating municipal, industrial, and hazardous wastes.

Oil/Water separator and recovery is used at some sites where free product exists in the groundwater. Undissolved liquid-phase organics are removed from subsurface formations, either by active methods (e.g., pumping) or a passive collection system. In an active pumping system, the combined oil and water is pumped above ground and passed through an oil/water separator, which separates out and skims the oil and stores it in an above ground container. This process is used primarily in cases where a fuel hydrocarbon lens is floating on the water table. Following recovery, it can be disposed of, re-used directly in an operation not requiring high-purity materials, or purified prior to reuse. Some of these systems may be designed to recover only product, mixed product and water, or separate streams of product and water.

3.2.4 Properties and behavior of inorganics (Metals)

Inorganics (metals) are also found as contaminants in groundwater. In the case of inorganics, specific technologies may often be ruled out, or the list of potential technologies may be narrowed, on the basis of the presence or absence of one or more of the chemical groups. Metals may be found sometimes in the elemental form, but more often they are found as salts mixed in the soil. At the present time, treatment options for radioactive materials are probably limited to volume reduction/concentration and immobilization. Asbestos fibers require special care to prevent their escape during handling and disposal; permanent containment must be provided. The properties and behavior of specific inorganic (metal) contaminants are discussed below.

Metals: Unlike the hazardous organic constituents, metals cannot be degraded or readily detoxified. The presence of metals among wastes can pose a long-term environmental hazard. The fate of the metal depends on its physical and chemical properties, the associated waste matrix, and the soil. Significant downward transportation of metals from the soil surface occurs when the metal retention capacity of the soil is overloaded, or when metals are solubilized (e.g., by low pH). As the concentration of metals exceeds the ability of the soil to retain them, the metals will travel downward with the leaching waters. Surface transport through dust and erosion of soils are common transport mechanisms. The extent of vertical contamination intimately relates to the soil solution and surface chemistry.

The properties and behavior of specific metals are discussed below:

Arsenic: Arsenic (As) exists in the soil environment as arsenate, As(V), or as arsenite, As(III). Both are toxic; however, arsenite is the more toxic form, and arsenate is the most common form. (Note: Arsenic is not a true metal; however, it is included here as it is considered as one of the hazardous metals in most of the developing countries.)

The behavior of arsenate in soil seems analogous to that of phosphate because of their chemical similarity. Like phosphate, arsenate is fixed to soil, and thus is relatively immobile. Iron (Fe), aluminum (Al), and calcium (Ca) influence this fixation by forming insoluble complexes with arsenate. The presence of iron in soil is most effective in controlling arsenate's mobility. Arsenite compounds are 4–10 times more soluble than arsenate compounds.

The adsorption of arsenite is also strongly pH-dependent. One study found increased adsorption of As(III) by two clays over the pH range of 3–9, while another study found the maximum adsorption of As(III) by iron oxide occurred at pH 7.

Under anaerobic conditions, arsenate may be reduced to arsenite. Arsenite is more subject to leaching because of its higher solubility.

Chromium: Chromium (Cr) can exist in soil in three forms: the trivalent Cr(III) form, Cr^{+3}, and the hexavalent Cr(VI) forms, $(Cr_2O_7)^{-2}$ and $(CrO_4)^{-2}$. Hexavalent chromium is the major chromium species used in industry; wood preservatives commonly contain chromic acid, a Cr(VI) oxide. The two forms of hexavalent chromium are pH dependent; hexavalent chromium as a chromate ion $(CrO_4)^{-2}$ predominates above a pH of 6; dichromate ion $(Cr_2O_7)^{-2}$ predominates below a pH of 6. The dichromate ions present a greater health hazard than chromate ions, and both Cr(VI) ions are more toxic than Cr(III) ions.

Because of its anionic nature, Cr(VI) associates only with soil surfaces at positively charged exchange sites, the number of which decrease with increasing soil pH. Iron and aluminum oxide surfaces adsorb the chromate ion at an acidic or neutral pH.

Chromium (III) is the stable form of chromium in soil. Cr(III) hydroxy compounds precipitate at pH 4.5 and complete precipitation of the hydroxy species occurs at pH 5.5. In contrast to Cr(VI), Cr(III) is relatively immobile in soil. Chromium (III) does, however, form complexes with soluble organic ligands, which may increase its mobility.

Regardless of pH and redox potential, most Cr(VI) in soil is reduced to Cr(III). Soil organic matter and Fe(II) minerals donate the electrons in this reaction. The reduction reaction in the presence of organic matter proceeds at a slow rate under normal environmental pH and temperatures, but the rate of reaction increases with decreasing soil pH.

Copper: Soil retains copper (Cu) through exchange and specific adsorption. Copper adsorbs to most soil constituents more strongly than any other toxic metal, except lead (Pb). Copper, however, has a high affinity to soluble organic ligands; the formation of these complexes may greatly increase its mobility in soil.

Lead: Lead is a heavy metal that exists in three oxidation states: O, +2(II), and +4(IV). Lead is generally the most widespread and concentrated contaminant present at a lead battery recycling site (i.e., battery breaker or secondary lead smelter).

Lead tends to accumulate in the soil surface, usually within 3–5 cm of the surface. Concentrations decrease with depth. Insoluble lead sulfide is typically immobile in soil as long as reducing conditions are maintained. Lead can also be biomethylated, forming tetramethyl and tetraethyl lead. These compounds may enter the atmosphere by volatilization.

The capacity of soil to adsorb lead increases with pH, cation exchange capacity, organic carbon content, soil/water Eh (redox potential), and phosphate levels. Lead exhibits a high degree of adsorption on clay-rich soil. Only a small percent of the total lead is leachable; the major portion is usually solid or adsorbed onto soil particles. Surface runoff, which can transport soil particles containing adsorbed lead, facilitates migration and subsequent desorption from contaminated soils. On the other hand, groundwater (typically low in suspended soils and leachable lead salts) does not normally create a major pathway for lead migration. Lead compounds are soluble at low pH and at high pH, such as those induced by solidification/stabilization treatment. Several other metals are also amphoteric, which strongly affects leaching. If battery breaking activities have occurred on-site, and the battery acid was disposed of on-site, elevated concentrations of lead and other metals may have migrated to groundwater.

Mercury: In soils and surface waters, volatile forms (e.g., metallic mercury and dimethylmercury) evaporate into the atmosphere, whereas solid forms partition into particulates. Mercury exists primarily in the mercuric and mercurous forms as a number of complexes with varying water solubilities. In soils and sediments, sorption is one of the most important controlling pathways for the removal of mercury from solution; sorption usually increases with increasing pH. Other removal mechanisms include flocculation, co-precipitation with sulfides, and organic complexation. Mercury is strongly sorbed to humic materials. Inorganic mercury sorbed to soils is not readily desorbed; therefore, freshwater and marine sediments are important repositories for inorganic mercury.

Zinc: Clay carbonates, or hydrous oxides, readily adsorb zinc (Zn). The greatest percentage of total zinc in polluted soil and sediment is associated with iron (Fe) and manganese (Mn) oxides. Rainfall removes zinc from soil because the zinc compounds are highly soluble. As with all cationic metals, zinc adsorption increases with pH. Zinc hydrolyzes at a $pH > 7.7$. These hydrolyzed species strongly adsorb to soil surfaces. Zinc forms complexes with inorganic and organic ligands, which will affect its adsorption reactions with the soil surface.

Radionuclides: For the purposes of remediation, radionuclides should be considered to have properties similar to those of other heavy metals. (See Table 3.1 for a list of typical radionuclides.) This does not imply that all radionuclides are heavy metals, but that the majority of sites requiring remediation of radioactively contaminated materials are contaminated with radionuclides that have similar properties. Like metals, the contaminants of concern are typically non-volatile and less soluble in water than some other contaminants. However, the solubility and volatility of individual radionuclides will vary and should be evaluated for each waste-stream being remediated. For example, cesium-137 is more volatile than uranium-238, and some cesium may volatilize, requiring off-gas treatment, when treated with processes at elevated temperatures (e.g., vitrification). Similarly, the mobility of radium-226, which is generally soluble in water under ambient conditions, will be greater than that of thorium-230, which is much less soluble.

Unlike organic contaminants (and similar to metals), radionuclides cannot be destroyed or degraded; therefore, remediation technologies applicable to radionuclides involve separation, concentration/volume reduction, and/or immobilization (Dutta, 1990). Some special considerations when remediating sites contaminated with radionuclides include the following:

Implementation of remediation technologies should consider the potential for radiological exposure (internal and external). The degree of hazard is based on the radionuclide(s) present and the type and energy of radiation emitted (i.e., alpha particles, beta particles, gamma radiation, and neutron radiation). The design should take into account exposure considerations and the principles of keeping exposures as low as reasonably achievable (ALARA).

Because radionuclides are not destroyed, ex situ techniques will require eventual disposal of residual radioactive wastes. These waste forms must meet disposal site waste acceptance criteria.

There are different disposal requirements associated with different types of radioactive waste. Remediation technologies addressed in this document are generally applicable for low-level radioactive waste (LLW), transuranic waste (TRU), and/or uranium mill tailings. The technologies are not applicable to spent nuclear fuel and, for the most part, are not applicable for high-level radioactive waste.

Some remediation technologies result in the concentration of radionuclides. By concentrating radionuclides, it is possible to change the classification of the waste, which impacts requirements for disposal. For example, concentrating radionuclides could result in LLW becoming TRU waste (if TRU radionuclides were concentrated to greater than 100 nanocuries/gm). Also, LLW classifications (e.g., Class A, B, or C for commercial LLW) could change due to the concentration of radionuclides. Waste classification requirements, for disposal of residual waste (if applicable), should be considered when evaluating remediation technologies.

Disposal capacity for radioactive and mixed waste is limited. For example, in the US commercial LLW disposal capacity will no longer be available for many out-of-compact (regions without a licensed LLW disposal facility) generator because the disposal facility in Barnwell, SC, closed (to out-of-compact generators) on 30 June 1994. Currently, there is only one disposal facility (Envirocare of Utah, Inc.) licensed to accept mixed waste (i.e., low-activity mixed LLW and hazardous waste) for disposal. Mixed waste can be treated to address the hazardous characteristics of the soil, thereby allowing the waste to be addressed as solely a radioactive waste.

3.2.4.1 Common treatment technologies for inorganics in groundwater

In addition to the general data requirements discussed earlier, it may be necessary to know other subsurface information to remediate inorganics in groundwater. Sometimes, treatability studies are necessary to ensure that the contaminated groundwater can be treated effectively at the design flow rate under a wide range of site-specific variability during the full-scale treatment process. A subsurface geologic characterization would also be important to evaluate the effects of adsorption and other processes of attenuation. In the case of large aquifers or complicated hydrogeologic characteristics, generally groundwater models are often needed to predict flow characteristics, changes in contaminant mixes and concentrations, and times to reach action levels.

Precipitation, filtration, and ion exchange are widely used ex situ treatment technologies for inorganics (metals) in groundwater and are discussed in the following paragraphs. In situ treatment technologies are used less frequently.

The combination of precipitation/flocculation and sedimentation is a well-established technology for metals and radionuclides removal from groundwater. This technology pumps groundwater through extraction wells and then treats it to precipitate lead and other heavy metals. Typical removal of metals employs precipitation with hydroxides, carbonates, or sulfides. Hydroxide precipitation with lime or sodium hydroxide is the most common choice. In general, the precipitating agent is added to water in a rapid-mixing tank along with flocculating agents such as alum, lime, and/or various iron salts. This mixture then flows to a flocculation chamber that agglomerates particles, which are then separated from the liquid phase in a sedimentation chamber. Other physical processes, such as filtration, may follow. A typical application of a precipitation/flocculation process designed to remove hexavalent chromium from contaminated groundwater at a site in western US is provided (Dutta, 1990) under reference at the end of this chapter.

Metal sulfides exhibit significantly lower solubility than their hydroxide counterparts, achieve more complete precipitation, and provide stability over a broad pH range. At a pH of 4.5, sulfide precipitation can achieve the EPA-recommended standard for potable water. Sulfide precipitation, however, can be considerably more expensive than hydroxide precipitation, as a result of higher chemical costs, increased sludge generation, and increased process complexity. Also, there are safety concerns associated with the possibility of H_2S emissions. The precipitated metals would be handled in a manner similar to contaminated soils. The supernatant would be discharged to a nearby stream, a publicly owned treatment system, or recharged to upstream of site aquifer. Selection of the most suitable precipitant or flocculent, optimum pH, rapid mix requirements, and most efficient dosages are determined through laboratory jar test studies.

Filtration isolates solid particles by running a fluid stream through a porous medium. The driving force is either gravity or a pressure differential across the filtration medium. Pressure-differentiated filtration techniques include separation by centrifugal force, vacuum, or positive pressure. The chemicals are not destroyed; they are merely concentrated, making reclamation possible. Parallel installation of double filters is recommended, so groundwater extraction or injection pumps do not have to stop operating when filters are backwashed.

Ion exchange is a process whereby the toxic ions are removed from the aqueous phase in an exchange with relatively innocuous ions (e.g., NaCl) held by the ion exchange material. Modern ion exchange resins consist of synthetic organic materials containing ionic functional groups to which exchangeable ions are attached. These synthetic resins are structurally stable and exhibit a high exchange capacity. They can be tailored to show selectivity towards specific ions. The exchange reaction is reversible and concentration-dependent; the exchange resins are regenerable for reuse. The regeneration step leads to a 2%–10% wastestream that must be treated separately.

All metallic elements present as soluble species, either anionic or cationic, can be removed by ion exchange. A practical influent upper concentration limit for ion exchange is about 2,000 mg/L. A higher concentration results in rapid exhaustion of the resin and inordinately high regeneration costs.

General effectiveness and cost of various treatment technologies for groundwater are summarized in Table 3.2.

3.3 TREATMENT TECHNOLOGIES FOR MUNICIPAL AND INDUSTRIAL WASTE – SOIL, SEDIMENTS, AND SLUDGES

It would be a monumental undertaking to provide details of each and every existing treatment technology for soil, sediments, and sludge. To simplify the effort, technologies have been grouped on the basis of common features and similarities in their ability to treat contaminants. The result of this grouping has yielded the following seven technology categories:

Table 3.2 List of groundwater treatment technologies – their efficacy and applicability

Technology	*Contaminant type*	*Efficacy*[a]	*Relative cost*[b]	*Maintenance/ reliability*	*Cleanup time (years)*	*Comments*
Ex situ treatments air stripping	VOCs	High	Low (≤$5.00/ 1,000 gallons)	High reliability, low maintenance	3–30	(USEPA, 1994b)
	SVOCs	Medium	Medium (>$8.50/ 1,000 gallons)	High	>30	
	Inorganics	Not Applicable (N/A)	N/A	N/A	N/A	
Carbon adsorption	VOCs	High	Medium	Medium reliability and maintenance	1–5	Usually used for low volume
	SVOCs	High	Medium	Medium	1–5	
	Inorganics	Low (seldom used)	Medium	Low Reliability	1–5	
UV oxidation	VOCs	High	High(>$8/ 1,000 gallons)	High Reliability medium maintenance	3–30	can treat liquid wastes from hospitals - nano-scale virus (e.g., COVID-19) particles.
	SVOCs	Medium	High	- Do -	5–40	
	Inorganics	N/A	N/A	N/A	N/A	Can not treat/remove inorganics(metals)
Natural attenuation	VOCs/SVOCs	Medium	Low	Low Reliability, Low Maintenance	Long (>30)	Used at sites with low concentrations or that are risky to pump and treat above ground
Bioreactors	VOCs/SVOCs	Medium	Low	High reliability, Low maintenance	2–20	Cannot be used at sites with very high concentrations.
Oil/water separators	Fuel	High	Low	High reliability, low maintenance	1–5	Usually applied to sites with small and confined volume/spill

(*Continued*)

Table 3.2 (Continued) List of groundwater treatment technologies – their efficacy and applicability

Technology	*Contaminant type*	*Efficacy*[a]	*Relative cost*[b]	*Maintenance/ reliability*	*Cleanup time (years)*	*Comments*
Precipitation/ flocculation	Inorganics (Metals)	High	High	High reliability, high maintenance	2–20	Applied only to remove metal contaminants
Filtration	Inorganics (Metals)	Medium	High	Low reliability, high maintenance	2–10	Often used as a tertiary or polishing step
Ion exchange	Inorganics (Metals)	Medium	High	Low reliability, high maintenance	2–20	Often used for small volumes or as a polishing step.
In situ treatment biological treatment	VOCs/SVOCs	High	Low	High Reliability, low maintenance	3–30	Generally suitable for low concentration of highly toxic organics. See References USEPA (1994a) and (2020).
Air sparging	VOCs	High	Medium	Medium reliability and maintenance	2–30	Generally suitable for VOCs only
Steam injection vapor extraction	VOCs/SVOCs	High	High	High reliability, high maintenance	3–30	Generally suitable for heavier and persistent contaminants. Not applicable to metals.
Slurry walls	VOCs/SVOCs or Inorganics	Medium	Very High	Low reliability, low maintenance	5–50	Slurry Walls Could leak underground.
Vacuum vapor extraction	VOCs	High	Low	High reliability, low maintenance	3–30	Works well with VOCs only
	SVOCs	Low	High	Low reliability	5–50	Not applicable to most SVOCs

[a] Efficacy: High: Contaminant can be treated/ removed over 99% or below acceptable concentration level; Medium: Contaminant cannot be treated/ removed over 90% or below the acceptable level; Low: Contaminant cannot be treated/ removed over 60% or below 10 times the acceptable level.

[b] Cost shown are in present (2020) U.S. Dollars ($). Any other definitions inside the table at the first citation of terms.

- Containment technologies
- Soil washing
- Thermal treatment
- Vapor extraction
- Bioremediation
- Incineration
- Other physical/chemical treatments.

Details on each of these treatment technology categories are covered in individual chapters beginning with Chapter 4.

REFERENCES

Dutta, S. (1990) Case Study of the Cleanup at Tinker AFB, Oklahoma. *Proceedings of the 84th Annual Meeting of the Air & Waste Management Association*, Vancouver, BC, June 1991.

Dutta, S. (2002) *Environmental Treatment Technologies for Hazardous and Medical Wastes, Remedial Scope and Efficacy*. Tata McGraw Hill Publishing Company, New Delhi.

USEPA. (1993) *Augmented In situ Subsurface Bioremediation Process. Bio-Rem, Inc.* EPA RREL, Demonstration Bulletin, Cincinnati, OH. EPA/540/MR-93/527.

USEPA. (1994a) *Remediation Technologies Screening Matrix and Reference Guide*. Federal Remediation Technologies Roundtable, Washington, DC. EPA/542/B-94/013, October.

USEPA. (1994b) *Ex-Situ Anaerobic Bioremediation System, Dinoseb, J.R. Simplot Company*. EPA RREL, Demonstration Bulletin. EPA/540/MR-94/508.

USEPA. (2020) *Superfund Remedy Report*. 16th Ed. Office of Land and Emergency Management, Washington, DC. EPA-542-R-20-001, July.

Chapter 4

Containment technologies

After completing the environmental assessment of a hazardous waste site, the very first step in the subsequent selection of a remedy is to decide whether the contaminants could be easily removed or destroyed. If removal or destruction becomes too difficult, and cost-intensive, or if removal or destruction poses a high threat and/or exposure to human health, and the environment, the next best choice for remediating the site is to contain the hazardous waste in a safe and stable manner and prevent migration and/or hazard posed by the contaminant(s) of concern. Various containment technologies are used for such cleanup activities as addressed below (USEPA, 2020).

4.1 DEFINITION AND SCOPE OF CONTAINMENT TECHNOLOGIES

Containment technologies use physical barriers to retain, immobilize, or isolate contaminated media from the surrounding environment and to minimize migration of the contaminants without destroying them (USEPA, 1997).

Many types of containment technologies are currently being used for soil treatment or disposal. Considering their similarity in treatment characteristics, end result of treatment, and cross-media transfer potentials, the following treatment technologies are listed as a few examples of containment technologies:

- Storage piles/Vapor sheds
- Storage containers/Drums
- Tank installations
- Impoundments
- Siltation basins
- Slurry walls
- Cement-based barriers
- Geomembrane barriers
- Frozen soil barriers
- Landfill cover systems
- Landfill liner systems
- Solidification and stabilization.

The scope of containment technologies is not limited to the above-listed treatment systems. Any treatment technology that meets the definition and scope of containment technologies

DOI: 10.1201/9781003004066-4

should generally be considered as a containment technology for its remedial applications, scope, and efficacy.

Containment of the contaminant source on-site primarily includes caps and cover systems. According to USEPA's recent Superfund site remedy analysis, more than 50% of the remedies selected used multiple remedial approaches, including various combinations of treatment, on-site containment, or off-site disposal, including institutional controls (ICs) such as restricting access to the area, while monitoring natural attenuation (USEPA, 2020).

A major challenge with the permanent storage of waste in a landfill is the tendency for toxic materials to move out of the landfill and into surrounding soils. This may occur in older unlined landfills through leakage and in both newer and older landfills through leaching, where water (from groundwater or from precipitation) enters the landfill and carries waste out. Both active and passive control measures may be used in the containment of a contaminant source. Modification of hydraulic gradients to regulate the direction of groundwater flow is an example of an active control, while passive control may involve the installation of a physical barrier. Both methods involve restriction in the movement of the contaminant away from the source area. Active and passive control measures can be used concurrently at some sites. Containment technologies thus prevent contamination from becoming more widespread, and as well allow for in situ treatment, i.e., treatment in place or original environment, and offer the potential for application of cost-effective innovative technologies that are not yet proven but nearing their commercial application stage. A successful containment system should minimize the potential hazards posed to human health and the environment to an acceptable risk level. Contaminant migration is controlled by isolating the contaminant source with the implementation of one of the containment technologies that best suits the site-specific conditions.

4.2 SPECIFIC CONTAINMENT TECHNOLOGIES

4.2.1 Solidification and stabilization

One method of preventing waste migration is through the use of waste solidification and/or containment. Two commonly used stabilization techniques are cement and organic polymer stabilization. In cement stabilization, a slurry of waste and water is mixed with standard industrial cement to form a solid cement block. In organic polymer stabilization, dried waste and prepolymer/catalyst are mixed. This mixture hardens into a sponge-like polymer. Landfill waste can also be contained with or without stabilization. In waste containment, injection wells are drilled, and grouting cement is pumped into the soils through the well casing or piping into the areas around the landfill. This process creates underground "walls" that stabilize the landfill boundaries and reduces the possibility of contaminants moving into the surrounding soils.

Solidification and stabilization provide a relatively quick way at a lower cost to prevent exposure to contaminants, particularly metals and radioactive contaminants. Solidification and stabilization have been selected or are being used in cleanups at over 250 Superfund sites across the country (USEPA, 2012a).

There are several advantages of cement stabilization:

- It is relatively low cost because of the use of readily available concrete and concrete mixing equipment.

- Concrete behavior is well understood, so the stabilization process is fairly simple.
- Concrete is also fairly versatile. It can be used to stabilize contaminants that contain heavy metals and since it is naturally alkaline, it can be used to neutralize/solidify acidic wastes.

The advantages of organic polymer stabilization include:

- Compatibility with wastes containing insoluble solids,
- Production of a non-flammable product, and less material as compared to cement stabilization. This results in a final product that is less dense than cement,
- Lower weight decreases transportation costs, storage cost, etc.

The advantages of waste containment by grouting include the isolation of the waste and allowing degradation of the contaminants in an isolated environment.

Since containment is a baseline technology, the limitations of the technology with the technical challenges are succinctly covered here. In cement stabilization, solids contained in the wastes are suspended, not chemically bound and are therefore subject to leaching. Water present around the final solidified waste form could carry certain types of contaminants out of the waste and into the surrounding environment. Also, the use of cement increases the waste volume, which still must be stored in a secure location. Finally, cement does not work with certain types of wastes such as organics, sodium salts, silts, clays, coals, lignite, etc. In organic polymer stabilization, solids contained in the wastes are also suspended and not chemically bound and are therefore subject to leaching. Also, with polymers, the resulting leach water that infiltrates through the polymer is strongly acidic, which can further harm the environment. The process to create a polymer requires experienced operators and special equipment. Toxic vapors may be produced during that process, and some resins are biodegradable and may decompose with time. Finally, "weep" water is produced as the polymer cures/hardens. This by-product is also strongly acidic in some cases and may contain toxic chemicals. In containment by grouting, the grouting process may only fill 30%–60% of the volume needing to be filled, which would still allow some leaching to occur. Therefore, careful monitoring of the injection process is necessary to ensure adequate grout flow rates and pressure to prevent a "breakthrough" of the grout to the surface. Many grouts must be pumped into the soil within a 6- to 8-hour period before it gets thickened and the flow is prevented.

4.2.2 Vertical barriers (slurry walls) and landfill covers/caps

Many final remedies at highly polluted sites, commonly referred to as Superfund sites in the US, have specified vertical barriers or cutoff walls as a component of the overall site remediation plan. To prevent the flow of clean/uncontaminated groundwater into the contaminated region, barrier walls can be placed upgradient from a contaminated site. This also helps in reducing the volume of contaminated water to be treated during the remedial process.

Although many different technologies fall within the umbrella of the containment technology group, the two most commonly encountered at present are slurry walls and landfills. A brief description of slurry walls and landfill cover systems is provided here.

These technologies often require extensive preparation of the site, e.g., geotechnical characterization and reinforcement of existing subsurface structures at the site before constructing a slurry wall. Possible pretreatment of the waste being contained, e.g., excavation and removal of highly contaminated waste from the site before constructing a landfill cover system (USEPA, 1997).

Slurry walls or the vertical engineered barriers (VEBs) may be selected at sites where cleanup of contaminated groundwater is difficult and expensive, or infeasible due to contaminant migration to areas where people and wildlife can come in contact with it before completion of cleanup. VEBs are also helpful in cases where cleanup methods could push contaminants to uncontaminated areas. In large contaminated areas, VEBs are typically less expensive to build and maintain than other types of technologies. VEBs have been selected or are being used at a number of Superfund sites across the US (USEPA, 2012b).

Where possible, the bottom of the VEB is "keyed into" a low-permeability layer of soil or bedrock. This means the bottom of the wall extends several inches into the soil or to the top of the bedrock, which helps to keep groundwater from seeping beneath the wall. A protective cap may be installed atop the VEB to prevent damage from vehicle traffic or other activities. A larger impermeable cap is often placed over the entire contaminated area enclosed by the VEB to prevent rainwater and snow melt from entering it.

Even when surrounded by a VEB and cap, contaminated groundwater may build up in the isolated area or move outward through small openings in the VEB toward clean areas. To prevent this, wells may be drilled within the isolated area to pump out groundwater. Contaminated groundwater that has been pumped to the ground surface usually requires treatment.

The VEB, cap, and pumping wells are maintained and monitored to ensure the contaminated area remains isolated and that contaminated groundwater does not spread to clean areas. Figure 4.1 shows an illustration of a VEB containment system around a contaminated area.

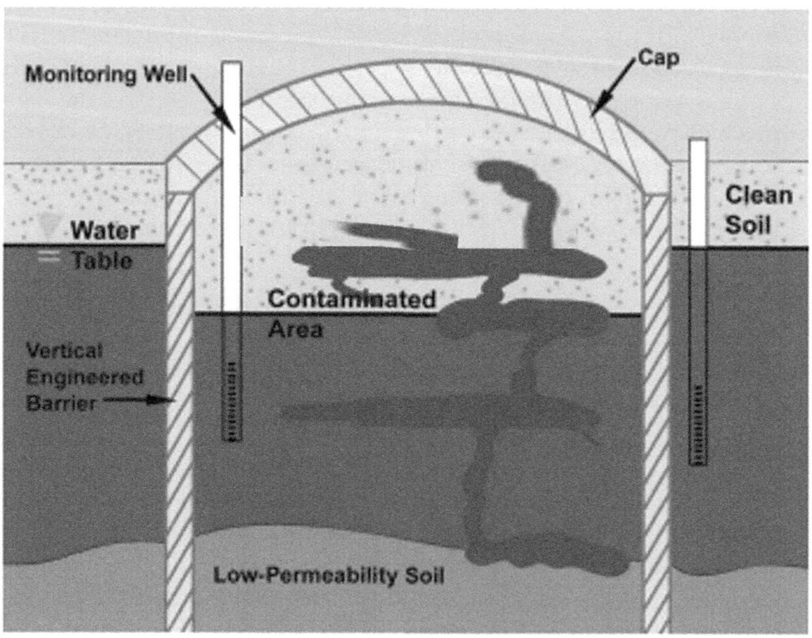

Figure 4.1 Typical slurry wall/VEBs (USEPA, 2012b).

The construction of slurry walls/VEBs involves the excavation of a vertical trench and using a bentonite-water slurry to hydraulically shore up the trench during construction followed by sealing the pores in the trench walls through formation of a "filter-cake". A cross-sectional view of a slurry wall is shown in Figure 4.1. Slurry walls are usually 20–80 feet deep with widths from 2 to 3 feet (USEPA, 1992a). Depending on the site conditions and contaminants, the trenches can be either excavated to a level below the water table to capture chemical "floaters" (this is termed as a "hanging wall") or extended ("keyed") into a lower confining layer (e.g., bedrock or aquitard). Similarly, on the horizontal plane, the slurry wall can be constructed around the entire perimeter of contaminated media or portions thereof (e.g., upgradient or downgradient). The principal distinctions among slurry walls are differences in the low-permeability materials used to backfill the trenches, namely the water content and ratios of bentonite/soil or bentonite/cement used to backfill the trench. In most cases, using bentonite/soil, the excavated soil is mixed with bentonite outside the trench. A relatively new development in the construction of slurry walls is the use of mixed in-place walls (also referred to as soil-mixed walls). This method of vertical barrier construction is recommended for sites where soft soils are encountered, there are concerns about the failure of traditional trenches due to hydraulic forces, or space availability for construction equipment is limited (USEPA, 1992a).

Grouting, including jet grouting, employs high-pressure injection of a low permeable substance into fractured or unconsolidated geologic material. This technology can be used to seal fractures in otherwise impermeable layers or construct vertical barriers in soil through the injection of grout into holes drilled at closely spaced intervals (i.e., grout curtain).

The design of landfill covers is also site-specific and depends on the intended functions of the system. Many natural, synthetic and composite materials and construction techniques are available. Covers can range from a one-layer system of vegetated soil to a complex multi-layer system. A cross-sectional view of a typical multilayer landfill cover is shown in Figure 4.2. In general, a fill layer of clean soils is placed first above the waste and graded to establish the base of the cover system. Then, a bottom layer, which may be a granular gas collection layer, is placed on top of the fill layer as a base for the remainder of the cover. The barrier layer is installed next. The materials used in the construction of the barrier layer are low-permeability soils and/or geosynthetic clay liners (GCLs). A flexible membrane liner (FML) layer is placed on top of the low-permeability barrier layer. These two layers prevent water infiltration into the waste. The high permeability drainage layer is placed on top of FML to drain the water

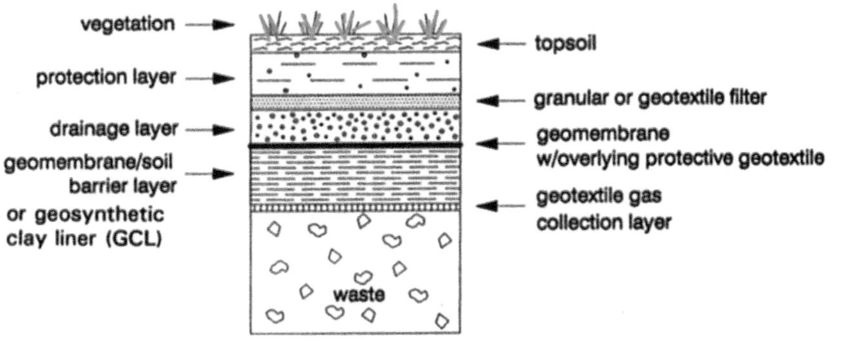

Figure 4.2 Engineered landfill cover system (USEPA, 1993a).

away that percolates through the top of the cover. A granular or geotextile filter fabric may be laid on top of the drainage layer to protect the drainage layer from clogging due to fine silts or clay deposits from the percolating water. A protective fill soil and topsoil are then applied and the topsoil seeded with grass or other vegetation adapted to local conditions. Covers are usually constructed in a crowned or domed shape with side slopes as low as is consistent with good stormwater runoff characteristics. Other materials may be used to increase slope stability. Steeply mounded landfills can have a negative effect on the construction and stability of cover. For example, there may be difficulty anchoring a geomembrane to prevent it from sliding along the interfaces of the geomembrane and soils.

Landfill covers are presently constructed in a variety of combinations depending upon the site-specific conditions. The most critical components of a cover with respect to the selection of materials are the barrier layer and the drainage layer. The barrier layer can be a GCL and/or low-permeability soil (clay). Other alternative barrier materials have also been identified in the document (USEPA, 1993a) cited under reference. Typical landfill cover is shown in Figure 4.2.

This typical landfill cover design is a well-established standard and practiced now in the second decade of the millennium. This typical landfill cover system design involves the following layers:

- A top layer consisting of two components:
  - either a vegetated or armored surface layer, and
  - a protection layer with a minimum thickness of 600 mm or 24 inches comprised of topsoil and/or fill soil with uniform slopes between 3% and 5%.
- A soil drainage layer (300 mm or 12 inches) with a minimum hydraulic conductivity of 1×10^{-2} cm/s or a geosynthetic drainage layer with equivalent performance characteristics and a final slope of at least 3%.
- A composite hydraulic barrier layer consists of:
  - an impermeable FML or geomembrane with a minimum thickness of 20 mil or 0.5 mm
  - a clay layer, 600 mm (24 inches) thick or a GCL with a maximum in-place saturated hydraulic conductivity (k) of 1×10^{-7} cm/s.

The diagrams of typical containment technologies are shown in Figures 4.1 and 4.2. Details of most of the above technologies are provided in the cited references (Dutta, 2002, USEPA, 1992a, 1993a). The salt bed disposal system used at the waste isolation pilot plant (WIPP) is a unique containment technology designed for the safe disposal of radioactive wastes (DOE, 1995).

A schematic cross-section of typical landfill caps or covers with the aforesaid layers is shown in Figure 4.3.

4.3 CROSS-MEDIA TRANSFER POTENTIAL AND BEST MANAGEMENT PRACTICES/OPTIONS (BMPs)

This section focuses on the identification of specific pathways for potential cross-media transfer of contaminants during the implementation of various containment technologies. Also, guidance on how to design and conduct remediation activities with minimal transfers of contaminants is provided here. Releases that may result in transfer of contaminants from the

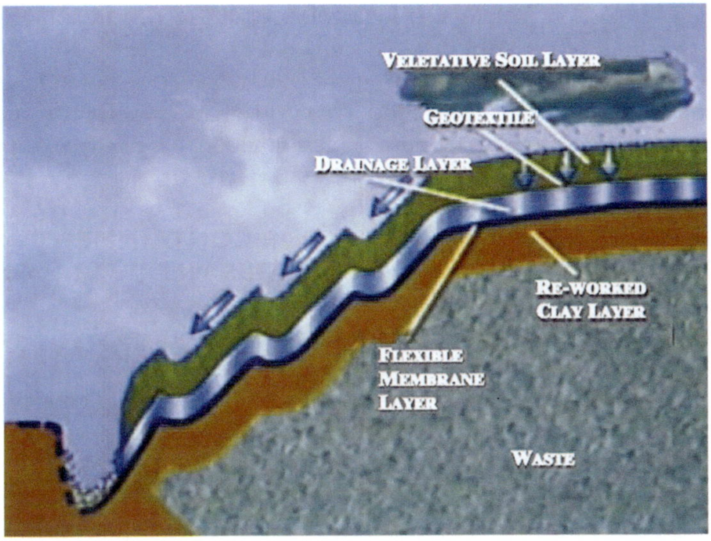

Figure 4.3 Schematic cross-section of typical landfill cap (Dutta, 2002).

soil or solid media to water, air or other natural media are generally referred to as cross-media transfer (USEPA, 1997). Various cross-media transfer potentials are addressed below.

4.3.1 General

The general cross-media transfer potential during site preparation, pre-treatment, and post-treatment activities are addressed later in Chapter 12 of this book.

4.3.2 Additional concerns and BMPs for specific containment technologies

In addition to the general concerns and BMPs that are addressed later in Chapter 12, containment technologies pose the following technology-specific concerns:

- Geomembranes are vulnerable to puncturing during installation. Inadequate preparation of the surface on which the geomembrane will be laid, or improper placement of materials on top of the geomembrane may result in punctures that allow infiltration of water and escape of volatile contaminants. Proper seaming of adjacent sheets also is critical for effective containment using this technology (Rumer and Ryan, 1995).
- Landfill cover systems pose the same cross-media transfer potential as geomembrane liners. Breaches in the system's integrity could allow infiltration of rainwater. The infiltration could then result in leaching of contaminants from the waste into surrounding soil and underlying groundwater. VOCs may also escape from the landfill cover system by diffusion through the cover layers and by "barometric pumping" through vents (USEPA,

1992c). Breaches in the landfill cover system and improper design and installation of landfill gas collection systems also could allow volatile contaminants to escape into the atmosphere.

- Since salt bed disposal is a deep underground entombment of contaminants, potential for releases is minimal. However, improper storage and handling of the wastes before placing them in the deep underground vaults could cause ground-level migration of contaminants.

4.4 BEST MANAGEMENT OPTIONS FOR CONTAINMENT TECHNOLOGIES TO CONTROL CROSS-MEDIA TRANSFERS

Technology-specific treatment activities and the possible BMP options to control cross-media transfer of contaminants during these activities are furnished below.

Containment treatment activities – During the implementation of the containment technologies, the following activities are most commonly undertaken:

- Excavation, trenching, storage of soils, sediments, and materials that will be used to construct containment system; construction of slurry walls, landfill covers, and other containment units. In the case of deep containment, such as with salt beds, extensive underground excavation is required. Secondary activities include surface water diversion and control, on-site pumping and treating, installation of cut-off trench type interceptors, and installation of leachate collection systems.

During these activities, the following BMPs should be considered for containment technologies:

- In the case of slurry walls, when there is a potential for outward migration and contamination of groundwater, periodic pumping and treating of the contaminants from the contained area and maintaining an inward hydraulic gradient from outside to inside the slurry wall may be considered. This practice has been observed in the field during field validation of BMPs (USEPA, 1997).
- All soils should be analyzed and processed before they are disposed of off-site.
- Air quality trends should be constantly monitored. If air quality degrades as a result of construction activities, those activities should be altered or stopped until air quality is restored.
- Climatological extremes (e.g., high wind) should be considered when implementing containment technologies.
- All debris should be covered during construction.
- Temporary sumps should be used to collect stormwater runoff from the site during construction.
- Temporary arrangements should be provided that protect areas that are vulnerable to damage and migration of contaminants during construction (e.g., road covers).
- Effective VOC, methane, and odor emissions should be controlled by using covers, foam suppressants, enclosures, vapor collection systems, gas flares, or other methods as appropriate.
- Contaminated liquids generated from treatment operations should be treated and/or disposed of protectively as specified later under general BMPs in Chapter 12.

- Additional post-treatment activities – In addition to the general post-treatment BMPs specified in Chapter 12, the following BMPs should be considered for containment technologies:
- Routine audits should be conducted to verify the integrity of the containment structure with accompanying documentation.
- For most containment technologies, the production of residuals is generally not a concern. However, the construction of soil-bentonite slurry walls can generate large quantities of excess slurry and excavated materials. In most cases, it is expected that these excess materials are not hazardous (Freeman and Harris, 1995). However, if the excavated soil and slurry cannot be used as backfill, they should be properly stored on-site or transported and disposed off-site.

4.5 WASTE CHARACTERISTICS IN CONTAINMENT TECHNOLOGIES THAT ARE LIKELY TO CAUSE CROSS-MEDIA CONTAMINATION

The effectiveness of containment technologies could be compromised and undue cross-media contamination caused under certain conditions, as provided below. However, it is possible to overcome some of these limitations with various technology-specific modifications and variations. Please refer to technology-specific references (Dutta, 1993, 2002; Francis and Spalding, 1991; USEPA, 1989, 1991b, 1992b) provided at the end of this chapter for additional information on modifications or variations that can be used to enhance the effectiveness of containment technologies.

- When contaminant concentrations exceed 10%–25% of their explosive limits, they are potentially unsafe to handle or can pose a threat to the integrity of the containment system (USEPA, 1994).
- Ignitable wastes may present fire hazards when treated using containment technologies that are exothermic (generating heat).
- Very low or high pH values – acidic (<4.0 Standard Unit (SU)) or alkaline, high (>11 SU) – may lead to corrosion in liners and equipment (USEPA, 1994).
- Strong oxidizers can corrode slurry walls and geomembranes.

4.6 TECHNICAL EFFECTIVENESS OF CONTAINMENT TECHNOLOGIES

Performance criteria:

A few frequently asked questions (FAQs) about the effectiveness of containment technologies are addressed below:

- *What contamination could remain after the technology is applied?*

  There is a significant reduction in contaminant mobility and toxicity with solidification. A significant mobility reduction is achieved with containment. In solidification, the majority of toxic chemicals in the landfill wastes are bound up in the hardened cement or polymer; therefore, the waste is much less likely to travel outside the hardened mass, rendering it to be much less toxic. However, even with stabilization, groundwater or precipitation can leach some contaminants out of the mass. Also, the overall volume of the waste, in many cases, increases by solidification – the volume increases by a factor of two (cement). Waste containment can provide a significant mobility reduction in landfill wastes. However, grout installation must be carefully conducted and the landfill carefully monitored to ensure adequate "walls" have been created around the landfill waste.

- *What process waste (secondary waste) does the technology produce?*

  There is little or no process waste with cement stabilization or waste containment (Corbitt, 1990). However, the process of stabilization using organic polymers can produce toxic vapors and leached water as the polymer cures/hardens. This leached water is termed "weep water", which is possibly a strongly acidic by-product and may contain toxic chemicals.
- *What are the requirements for the treatment or storage of the secondary waste?*

  Further treatment of the process waste is not generally conducted. The secondary waste needs to be disposed of using the proper procedures and in compliance with the applicable local, state, and federal regulations.

The requirements for decontamination or decommissioning of equipment are as follows:

Solidification may require decontamination of the mixers used to combine the waste with the cement or polymer used in the process. Waste containment may require decontamination of the drilling rigs and the grout pumping equipment. Standard site decontamination procedures can be used in these cases.

For disposal of the secondary waste, the following possibilities should be considered:

Solidification will require a final disposal site for the waste/cement or waste/polymer mixture. Stabilization leaves the waste in the original landfill, so additional disposal is not needed. Some polymers may produce "weep water" during and after the curing process. This water may be highly acidic and may contain components of the original contaminants. Hence, during the final disposal of this waste, precautions must be taken to guard against any adverse impact of the decontamination process on the environment. Such seepage or leaching may also occur through the use of cement and polymers.

4.7 PRACTICABILITY

While considering the practicability of the chosen containment technology option, one may also consider the cleanup options that are likely to be precluded by this technology in the future.

For example, solidification would foreclose many future options for the waste management. It would be very difficult to separate the waste from the cement or polymer in the future. Containment, on the other hand, would not foreclose future option for the waste. The waste could be removed from the landfill in the future, and the grouts used for containment could be removed with heavy equipment. Containment can also be used as an interim measure, while other options for waste treatment are examined.

A few important questions concerning the containment technology are as follows:

- *How reliable is the technology and what controls are available to minimize environmental damage or migration of contaminants in case this technology fails?*

  The reliability of solidification depends upon the type of waste. For cement-based solidification, organics, sodium salts (of arsenate, borate, phosphate, etc.), and large amounts of sulfates in the waste can prevent or severely retard hardening. Both cement and polymer stabilization techniques form suspended waste masses which are subject to leaching. Containment may only fill only 30%–60%of the required volume. Therefore, leaching can still occur.

Adequate testing must be done before solidification is attempted. If the final mixture does not harden, the waste disposal issue becomes much larger and a more complicated problem in the form of a large mass of wet cement and waste that would have to be disposed of with utmost care.

Grout installation must be carefully conducted and the landfill carefully monitored to ensure that adequate "walls" have been created around the landfill waste.

- *Is it easy to use the technology?*

  Solidification is fairly easy to use. There is extensive information available on various containment technologies. Also, some experienced individuals are available from the construction industry who can provide expertise in cement mixing and setting . Polymer solidification requires personnel experienced in the mixing and setting of the polymer being used. Also, proper safety precautions must be taken to protect workers against the toxic by-products of these activities.

  Containment technologies involve a lot of site-specific adaptations and require a close study of the soil and waste characteristics. The grout must be able to penetrate the soil volume in order to create an effective barrier that prevents waste leaching. Therefore, the grout must be carefully mixed and pumped under well-calculated pressures.
- *What infrastructure is needed to support the technology?*

  Cement mixers and heavy pumping equipment are involved in implementing this technology. Waste containment is a specialized process requiring specialized experience and equipment. It requires high-pressure grout pumping equipment and drilling of boreholes. It also requires the installation of appropriate instruments to verify barrier integrity.
- *How versatile is the technology?*

  Solidification works well with certain types of contaminants and not at all with others. For example, cement does not work with most organic wastes, sodium salts, silts, clays, coals, lignite, etc. Also, solidification is based on well-understood cement mixing and handling technology and the required equipment are widely available. Polymer solidification requires more experienced personnel and specialized equipment but requires less polymer and is able to produce a final waste form that is much lighter than the waste form produced by cement techniques. Containment is fairly versatile but must be performed under well-understood soil and waste conditions.

These technologies are somewhat stand-alone in requiring site-specific design for most cases.

- *Can the technology be procured "off the shelf"? Which components are available and which must be developed?*

Cement solidification is based on cement mixing and handling. This practice is well known, and widely available equipment are used for this purpose. Polymer solidification requires more specialized equipment, but that is still commercially available. Containment technologies still require specialized equipment, in most cases, that would have to be custom designed and implemented for the specific site. However, recently, many remediation companies specializing in grout-based containment have come forward and undertaken such projects effectively.

- *How difficult is it to maintain the equipment for this technology?*

Solidification equipment is easily maintainable since it is based on commercially available equipment. Containment requires specialized equipment and would have to be maintained by the developer.

- *What equipment safety measures are needed and what measures are in place to protect workers and the public?*

Care must be taken not to expose workers to hazards during the mixing process. In addition, protective equipment would be needed due to toxic by-products of polymer curing. In general, however, standard hazardous waste site safety procedures would provide adequate safety measures. Also, the best management practices (BMPs) options for the containment technologies as listed earlier in this chapter should be used for proper control of cross-media transfer of contaminants and protection of workers and the public.

4.8 COST FACTORS

Cement stabilization generally costs \$35–\$50 per cubic yard currently (2020) in the US. However, this will vary from location to location, and country to country (Dutta, 2002). This information is not available for organic polymers and for waste containment technologies. These technologies are used in special cases, for certain waste types or to provide containment for certain landfills. In addition, general start-up cost information is currently not available.

Once the waste has been stabilized and/or contained, the cost of the main operations would be for the site monitoring programs to ensure that no leakage or leaching is occurring. That cost is very minimal. However, the cost of groundwater monitoring could be high in certain cases. Due to the long life cycles of landfills – 30 years according to current regulations in the US – the total life cycle costs are not currently available.

The time required for the implementation of the containment technologies varies from technology to technology. Solidification usually requires less than 3 days – the time involved in mixing the cement/waste and hardening. Containment requires more time – usually several weeks to a total of few months for access road, site preparation, borehole drilling, grout injection, hardening time and other related activities.

4.9 WORKER SAFETY

Worker safety issues are also a major concern. In most containment technologies, worker hazards include exposure to hazardous waste at a landfill site, toxic "weep water" and fumes from polymer curing, and heavy equipment hazards during grout installation. However, these techniques are fairly standard at present and standard protective measures and clothing are sufficient. Standard heavy equipment are generally used to do the work involved including drill rigs, cement mixers, polymer curing equipment, grout pumps, etc. The number of people required to operate the technology also varies. Solidification would require a cement/polymer crew to operate the mixers, cure the polymer, etc. In general, this would require approximately eight people. Containment would require a drilling crew and grout pumping crew. This too would involve approximately eight people.

The history of containment technology applications and related accidents would provide an insight into its safety factors.

The chances for public accidents from these technologies are minimal. The public (and other workers) would not be in an area where these technologies are being applied. Cement solidification produces a very heavy mass, which is very difficult to transport; therefore, this technique would generally be applied near the final storage area for the waste. Polymer solidification generally requires secondary containment in steel drums. These steel drums are easier to transport but need to be specially stored and monitored for leakage.

4.10 ENVIRONMENTAL IMPACT

The impact of containment technologies on the ecology of the area is a matter of concern to the communities and local authorities. The solidified mass of either cement or polymer would still need a final storage location, which would have to be chosen specifically to minimize possible ecological impact. For example, waste could be solidified with cement and placed back into the original landfill from which it came. A landfill cap or cover could then be constructed to prevent water from infiltrating into the waste area and thus leaching of contaminants could be prevented. The combination of stabilization and capping would be much more effective than capping alone. Similarly, grout-based containment and capping would be much more effective. In addition, containment could be used to prevent leakage from older landfills until alternative methods of treating the waste could be developed. In most cases, where a landfill or storage area is needed, the area would probably be landscaped finally to make it aesthetically appealing.

The natural resources used in technology development, manufacture, or operation is another factor of consideration. Although cement is an energy-intensive material to produce, yet no special energy demand exists for its use.

4.11 LAND USE

Solidification might produce a site for future use. Containment could work as well, but the waste would still be in its original form – possibly hazardous. In general, the capacity for the unrestricted use of a site is a regulatory issue that is very much local and site-dependent.

There is a potential for economic impacts of using this technology, which has both a positive and a negative side to it. If a site could be shown to be "safe" after a stabilization or containment technology has been applied, the site might be reusable, providing a positive economic impact on an area. However, proper groundwater monitoring wells, both up- and down-gradient of the site and periodic air quality monitoring might have to be conducted to establish safe conditions for use by the local residents. Moreover, containment or stabilization of waste might produce a very permanent landfill site with an overall negative economic impact on the area.

Stabilization and containment technology implementations generally involve a fairly small labor force and are completed with no resulting long-term demands. Concerning the regulatory objectives and cleanup milestones, the implementation of these technologies depends heavily on the site involved and the site-specific issues, such as waste types, environmental impacts, and long-term use. These issues must be carefully considered to determine if these technologies would be compatible with applicable local regulations and the cleanup milestones.

REFERENCES

Corbitt, R.A. (1990) *Standard Handbook of Environmental Engineering*, pp. 9.36–9.43.

Dutta, S. (1993) Modified Cover System for Hazardous Waste Landfills in Semi-Arid Areas. *Proceedings of the 3rd International Conference on Case Histories in Geotechnical Engineering*, St. Louis, MO, June.

Dutta, S. (2002) *Environmental Treatment Technologies for Hazardous and Medical Wastes, Remedial Scope and Efficacy*. Tata McGraw Hill Publishing Company, New Delhi.

Francis, C.W., and B.P. Spalding. (1991) In Situ Grouting of Low-Level Burial Trenches with a Cement Based Grout. *Oak Ridge National Laboratory, proceeding of the conference: Environmental Remediation 1991 in Pasco*, Washington, DC.

Freeman, H.M., and E.F. Harris. (1995) *Hazardous Waste Remediation: Innovative Treatment Technologies*. Technomic Publishing Co., Inc., Lancaster, PA.

Fung, R. (ed.) (1990) *Protective Barriers for Containment of Toxic Materials*. Noyes Data Corporation, Park Ridge, NJ; 1980, Pollution Technology Review No. 66.

Rumer, R.R., and M.E. Ryan. (1995) *Barrier Containment Technologies for Environmental Remediation Applications*. John Wiley and Sons, New York, August.

U.S. Department of Energy (DOE). (1995) *Draft No Migration Variance Petition, Waste Isolation Pilot Plant*. Carlsbad Area Office, Carlsbad, NM. DOE/CAO-95-2043, May.

U.S. Environmental Protection Agency (USEPA). (1989) *Technical Guidance Document, Final Covers on Hazardous Waste Landfills and Surface Impoundments*. Office of Solid Waste and Emergency Response, Washington, DC. EPA/530/SW-89/047.

USEPA. (1991a) *Handbook, Stabilization Technologies for RCRA Corrective Actions*. Office of Research and Development, Washington, DC. EPA/625/6-91/026, August.

USEPA. (1991b) *SITE Technology Demonstration Summary, International Waste Technologies/Geo-Con In Situ Stabilization/Solidification Update Report*. Center for Environmental Research Information, Cincinnati, OH. EPA/540/S5-89/004a.

USEPA. (1992a) *Engineering Bulletin-Slurry Walls*. Office of Research and Development, Cincinnati, OH. EPA/540/S-92/008, October.

USEPA. (1992b) *Engineering Bulletin, Control of Air Emissions from Materials Handling During Remediation*. Office of Research and Development, Cincinnati, OH. EPA/540/2-91/023, October.

USEPA. (1992c) *Organic Air Emissions from Waste Management Facilities*. U.S. Environmental Protection Agency, Washington, DC. EPA/625/R-92/003, May.

USEPA. (1993a) *Engineering Bulletin-Landfill Covers*. Office of Research and Development, Cincinnati, OH. EPA/540/S-93/500, February.

USEPA. (1993b) *Environmental Fact Sheet: Controlling the Impacts of Remediation Activities in or Around Wetlands*. Office of Solid Waste and Emergency Response/Office of Waste Programs Enforcement, Washington, DC. EPA/530/F-93/020, August.

USEPA. (1994) *BMP Development Workshop Summary – Containment Technologies*. Office of Solid Waste, Permits and State Programs Division, Washington, DC, August.

USEPA. (1997) *Best Management Practices for Soil Treatment Technologies*. Office of Solid Waste, Washington, DC. EPA 530-R-97-007, May.

USEPA. (2012a) *A Citizen's Guide to Solidification and Stabilization*. Office of Solid Waste and Emergency Response, Washington, DC. EPA-542-F-12-019, September.

USEPA. (2012b) *A Citizen's Guide to Solidification and Stabilization*. Office of Solid Waste and Emergency Response, Washington, DC. EPA-542-F-12-022, September.

USEPA. (2020) *Superfund Remedy Report*. 16th Ed. Office of Land and Emergency Management, Washington, DC. EPA-542-R-20-001, July.

Chapter 5

Soil washing

The soil washing technology is primarily used to remove contaminants that are adsorbed or adhered to soils or solid media. Different soil washing technologies and processes are discussed in this chapter.

5.1 DEFINITION AND SCOPE OF SOIL WASHING

Soil washing is an innovative treatment technology, generally, a water-based process, sometimes combined with chemical additives and mechanical process to scrub soils and remove hazardous contaminants by extracting them into a smaller volume (USEPA, 2002). This technology is used to remove a wide range of contaminants such as organic, inorganic, and radioactive contaminants from soils or solid media (USEPA, 1993a and 2002). This aqueous-based technology uses mechanical processes (e.g., scouring) and/or solubility characteristics of contaminants to separate contaminants from excavated soils or solid media. The process frees and concentrates contaminants in a residual portion of the soil (typically 5%–40% of the original volume), where they can be subsequently treated by other remediation techniques or managed in compliance with applicable regulations (USEPA, 1997).

Many different types of soil washing treatment technologies are generally used for removing contaminants from soils. The following treatment technologies and processes are listed as a few examples of soil washing:

- Solvent extraction
- Debris washing
- Magnetic separation
- Froth flotation
- Excavating, dredging, and conveying
- Wet and dry screening
- Gravity concentration.

A typical schematic of a soil washing system is shown in Figure 5.1.

The scope of soil washing is not limited to the above-listed technologies. Any treatment technology that satisfies the key features of soil washing could be considered as soil washing technologies. Solvent extraction has been included in this chapter because the treatment process closely matches the key features of soil washing.

DOI: 10.1201/9781003004066-5

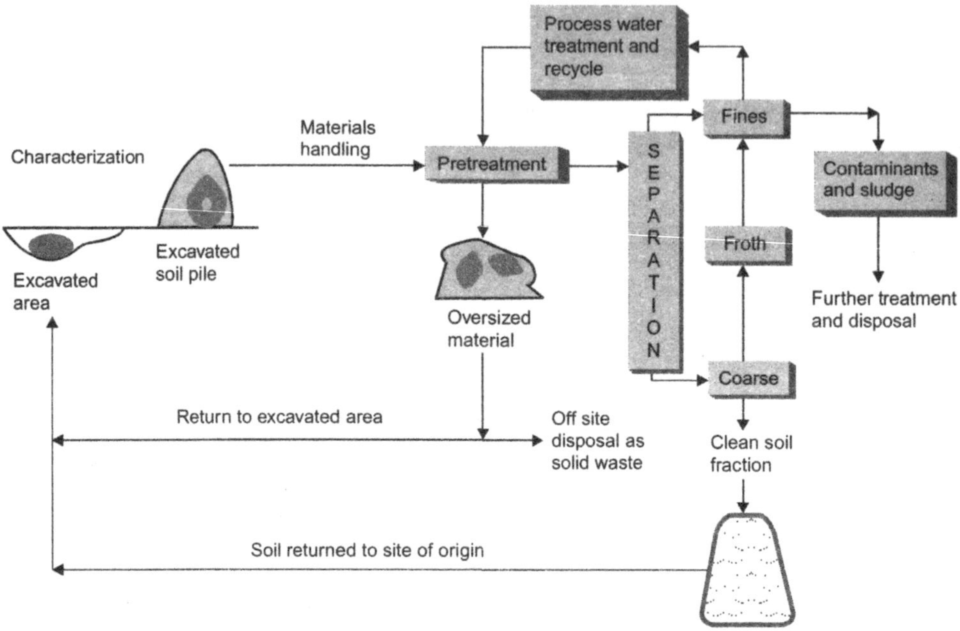

Figure 5.1 Basic soil washing flow diagram (Source: USEPA, 1993a)

5.1.1 Key features of soil washing technology

- A non-destructive process that separates contaminants from solids and concentrates the contaminants for collection and/or treatment.
- An ex situ technique normally requires excavation of soil or sediment and other materials handling operations, such as pre-screening of oversize (e.g., vegetation, debris, etc.), stockpiling, conveying, and particle size separation.
- Applicable for treating a wide variety of organic, inorganic, and radioactive contaminants in soil or solid media.
- Commonly relies on additives such as surfactants or solvents to enhance the effectiveness of the soil washing process.
- Significantly reduces the volume of contaminated soil (USEPA, 1992a).

5.2 SOIL WASHING TECHNOLOGY DESCRIPTION

A synopsis of the technology description is provided here. For detailed information on this technology, see the relevant references listed at the end of this chapter (USEPA, 1994a, 1994b, 1993b). Soil washing is a water-based process for scrubbing soils ex situ to remove contaminants (USEPA, 1994c). The process removes contaminants from soils in one of two ways:

- By dissolving or suspending them in the wash solution (which is later treated by conventional wastewater treatment methods).

- By concentrating them into a smaller volume of soil through particle size separation, gravity separation, and attrition scrubbing (similar to those techniques used in sand and gravel operations).

Soil washing systems incorporating most of the removal techniques offer great promise for application to soils contaminated with a wide variety of heavy metals, radionuclides, and organic contaminants.

The concept of reducing soil contamination through the use of particle size separation is based on the finding that most organic and inorganic contaminants tend to bind, either chemically or physically, to clay, silt, and organic soil particles. The silt and clay, in turn, are attached to sand and gravel particles by physical processes, primarily compaction and adhesion. Washing processes that separate the fine (small) clay and silt particles from the coarser sand and gravel soil particles effectively separate and concentrate the contaminants into a smaller volume of soil that can be further treated or disposed of. Gravity separation is effective for removing high or low specific gravity particles such as heavy metal-containing compounds (lead, radium oxide, etc.). Attrition scrubbing removes adherent contaminant films from coarser particles. The clean, larger fractions are generally found to be clean with no adsorption of contaminants. After separation, these clean fractions can be returned to the site for continued use.

The soil washing technology offers the potential for the recovery of metals and can clean a wide range of organic and inorganic contaminants from coarse-grained soils. The soil washing process begins with the excavation and preparation of the feedstock soil. Soil characteristics can vary widely within a relatively small area, and it is common to find that the top layer of soil in urban areas is composed of fill materials, not native soils (USEPA, 2002). However, local soil maps and other general soil information can be used for screening purposes. Soil preparation can involve the mechanical screening of the feedstock to remove rocks, debris, and other oversized materials. The treatment process generally involves the use of wet, mechanical scrubbing and screening processes to separate particles containing the contaminants. Most commercially available soil washing systems utilize mechanical screening devices to remove oversize materials and separation systems to generate coarse- and fine-grained fractions. The process also includes treatment units for washing and systems for scrubbing the separated fractions. The specific processes and equipment used depend upon individual site characteristics (USEPA, 1993a).

After excavation and preparation, the feedstock soil is actively mixed with water or an amended water-based washing fluid, which separates the contaminants from the soil. The soil is then separated from the spent fluid, and the soil is recovered in two distinct fractions. One fraction comprises a relatively high volume – coarse sand and gravel fraction that is clean and suitable for use as on-site fill; the other usually comprises a smaller volume, fine silt, and clay fraction that typically carry the bulk of the contaminants. From the coarse soil fraction, a contaminated, naturally occurring organic material may be separated as a third fraction by specific gravity separation. The coarse sand and gravel fraction are generally passed through an abrasive scouring or scrubbing action to remove the surficial contamination. The wash-water in this washing step may contain a basic leaching agent, surfactant, or chelating agent to help remove organics or heavy metals. The mixture is agitated by the use of high-pressure water jets, vibration devices, and other means depending upon the equipment (USEPA, 1991a). Fine particles are sometimes separated further in a sedimentation tank using a flocculating agent.

In the final step, the remaining fine silt, clay, and the contaminated wash-water are treated. The contaminated wash-water may require precipitation and clarification, which removes metals and fine soils as a sludge. The fine soils, in which contaminants have been concentrated, will normally require further treatment or proper disposal in compliance with applicable regulations.

To increase the efficiency of contaminant removal, sometimes chemical agents are added to the wash-water. Acids, such as hydrochloric acid, sulfuric acid, and nitric acid, may be added to improve the solubility of certain contaminants, especially heavy metals. Sodium hydroxide, sodium carbonate, and other bases can be used to precipitate contaminants in the extraction fluid. Dispersion of oily contaminants can be facilitated by the addition of surface-active agents. Various chelating agents such as citric acid, ammonium acetate, nitrilotriacetic acid (NTA), and ethylenediaminetetraacetic acid (EDTA) will remove the available fraction of inorganic contaminants (USEPA, 1991b). For improved removal in certain cases, the extraction temperature is elevated or an oxidizer, such as hydrogen peroxide, or ozone is added for chemical oxidation.

While soil washing is used extensively in Europe, it has had limited use in the US. It is sometimes used in combination with other soil treatment processes for achieving higher cleanup goals. During 1986–1989, the technology was one of the selected source control remedies at eight Superfund sites.

Soil washing is most commonly used in combination with the following technologies: bioremediation, incineration, and solidification/stabilization. Depending on the process used, the washing agent and soil fines are often found to exist as residuals that require further treatment. When contaminated fines have been separated, clean coarse-grain soil can usually be returned to the site. The time to complete the cleanup of a site using soil washing can be several months. For example, a "standard" 18,200-metric-ton (20,000-ton) site using soil washing could be typically 3–5 months.

The current (2020) average cost for use of this technology, including excavation, is approximately \$220–\$375 per metric ton (\$200–\$340 per ton) in the US, depending on the target waste quantity and concentration (Dutta, 2002). Information from a few sites where soil washing treatment has been used is furnished in Table 5.1.

5.3 AIR EMISSION AND PUBLIC PERCEPTION ISSUES INVOLVING SOIL WASHING TREATMENT TECHNOLOGY

As mentioned in Chapter 11, fugitive dust emission from soil washing treatment process has been one of the major issues raised by the neighboring communities at certain treatment sites where soil washing was used. However, many different control options or the best management practices (BMPs) are available for mitigating and suppressing fugitive dust emission as specified later in this chapter. The proper use of such BMPs will also help in building community alliance and support in cleanup actions involving soil washing treatments.

5.4 CROSS-MEDIA TRANSFER POTENTIAL OF SOIL WASHING TECHNOLOGIES

5.4.1 General

Cross-media transfer potentials during site preparation, pre-treatment, and post-treatment activities are addressed in Chapter 12.

Table 5.1 Site information on soil washing treatment (Dutta, 2002)

Site name	*Summary*	*Beginning levels*	*Levels attained*	*Costs in U.S. (2020)***
Toronto Port Industrial Dist. Ontario, Canada. Toronto Harbor. 60 Harbour St. Toronto, CA M5J 1B7.	Soil washing (volume reduction), metal dissolution, and chemical hydrolysis with biodegradation (organics)	52 ppm Naphthalene; 10 ppm benzo(a)-pyrene	<5; 2.6 ppm	NA
Montclair Superfund Site Montclair, NJ	Attrition mills, classifiers, and filter press to reduce the amount of low-level radioactive waste to be disposed of, 56% volume reduction.	NA	11 pCi/g	$510/hour of operation
Excalibur Technology. EPA RREL. 26 West M.L. King Dr., Cincinnati, OH 45268	Bench scale Soil washing and catalytic ozone oxidation. Site demo – Coleman Evans, Florida	20,000 ppm total capacity	NA	$156–$289/m^3 ($70–$221/yd^3)
Alaskan Battery Enterprises Superfund Site, Fairbanks, AK. EPA RREL 2890 Woodbridge Ave. Building 10, Edison, NJ	Pilot scale, featuring gravity separation and particle size classification	2,280–10,374 ppm lead	15–2,541 ppm	
Twin Cities AAP. New Brighton, MN	Full scale, featuring gravity separation, particle size classification, metal leaching, and lead recovery	NA	Targets reached state goals for Cr, Cu, Hg, and Ni.	NA

**Current (2020) cost figures are calculated on the basis of the costs listed in the reference (Dutta, 2002) for this table.

5.4.2 Additional concerns for soil washing technologies

- Additives used in the process can increase the potential of direct spillage of wastewater (e.g., foam with metals and organics) to the soil and surface water during treatment activities, especially if soil washing unit is not properly lined and bermed.
- In the specific case of solvent extraction, where there are pressurized tanks with highly flammable and volatile solvents are commonly used, there is potential for VOC emissions due to leaks in pipes, joints, and valves. Major emission points associated with solvent extraction are those involved in the distillation process used to recover the solvent (USEPA, 1992b).
- Chelating agents, surfactants, solvents, and other additives are often difficult and expensive to recover or recycle from the spent washing fluid by conventional treatment processes, such as settling, chemical precipitation, or activated carbon. The presence of additives in the contaminated soil and treatment sludge residuals may cause added difficulty in disposing of these residuals (USEPA, 1993a), thereby increasing the potential for cross-media transfer.
- Additives used in the soil washing, debris washing, wet screening, and froth flotation process can increase the potential of direct spillage of wastewater to the soil and surface water during treatment activities as a direct result of excessive foaming or frothing.
- Soil characterization data (e.g., size classifications, levels of contamination, permeability of soil, and estimates of soil quantities) used for treatability or pilot-scale tests may not accurately reflect the breadth of soil characteristics actually found in the field. Accurate characterization is important for the efficient use of this technology, and additional pretreatment of the soil (i.e., additional drying, crushing, and sizing) may be necessary just before operating the technology. Such improper characterization or lack of adequate pretreatment may lead to a higher potential for cross-media transfer than expected.
- Treated soil residues from soil washing, wet and dry screening, gravity concentration, and froth flotation may have significantly different soil characteristics such as permeability and compactability, and thus could adversely affect the groundwater flow characteristics of the site where these soils are replaced. Other constituents at the site could then migrate back into or through the treated soil.
- Wastewater from soil washing, wet screening, froth flotation, and debris washing may contain diluted amounts of the hazardous constituents and significant levels of suspended matter. Cross-media transfer can occur if these residues are released to the environment without any treatment.
- Improper or incomplete identification of contaminants and lack of knowledge with respect to their concentrations in the spent wash-water may foul up the system since the wash-water is treated and recycled back into the washing process. This may in turn cause inadequate cleaning and removal of contaminants and cause cross-media transfer.
- When the fine particles are not separated from the pre-treated soils, it may result in emission of the fine particles, which oftentimes bind most of the contaminants.

5.5 BEST MANAGEMENT OPTIONS FOR CONTROLLING CROSS-MEDIA TRANSFER POTENTIAL OF SOIL WASHING TECHNOLOGIES

General BMPs to prevent potential cross-media transfer of contaminants during pretreatment and post-treatment activities are discussed in Chapter 12. Furnished below are some technology-specific treatment activities and possible BMP options to control cross-media transfer of contaminants during these activities as well as a few post-treatment BMPs.

During the soil washing treatment process, the following activities are most commonly undertaken:

- Excavation of soils, temporary storage, particle size separation, transportation/transfer of contaminated soils from loaders to dump trucks, mixing action, movement of the contaminated media through a conveyor system, desorption, separation, and washing in an aqueous media.

The following BMPs, when appropriate, are recommended to prevent cross-media transfer of contaminants for the above activities during the soil washing process.

- Precautions should be taken to avoid foaming (or frothing) and subsequent overflow by periodically performing visual inspections when additives are used that have been demonstrated to froth in other situations. Field testing of small soil samples in jars with excess additives might help avoid anticipated problems but should not be used as the only means to predict frothing problems. As a contingency plan, the area underneath the soil washing unit could also be lined and bermed to collect any potential spillage.
- Major emission points associated with solvent extraction, such as those involved in the distillation process for recovering solvent, should be carefully monitored during operation. Process shutdowns may be deemed necessary if excessive levels of emissions are detected.
- Volumes of soil batches should be carefully managed so that they do not overfill the containers or exceed the normal operating specification of the equipment. The soil batches should preferably be run at less than maximum capacity to prevent leaks or spills.
- Chelating agents, surfactants, solvents, and other additives should be carefully selected to avoid ones that are difficult and expensive to recover or treat by conventional processes, such as settling, chemical precipitation, or activated carbon. The potentially adverse impact of residual soil washing additives that are anticipated to remain in the soil after treatment should be examined and replacement plans adjusted accordingly (e.g., if acids are used to extract metals from soils, the residual soil may need to be either limed prior to replacement in order to account for the acidity expected to be left in the soil or a neutralization step may need to be included as part of the soil washing process).
- Any off-site runoff should generally be prevented from entering and mixing with the on-site contaminated media by building earthen berms or adopting other similar measures. Provision should be made to capture the on-site surface water runoff by diverting it to a controlled depression-area or a lined pit.
- Most soil washing operations are vulnerable to high wind, especially due to the fugitive dust emission from these operations. Weather monitoring and operational control should be exercised as specified in Chapter 12 of this book. During excavation and material handling activities, meteorological conditions should be strongly emphasized and evaluated to minimize cross-media transfer.
- Mixing, crushing, or conveying activities should generally be conducted under an environment where the off gases, volatiles, dust, etc. are all captured inside a hood or cover or controlled using other control options listed in Chapter 13. The VOC emissions associated with these activities should be controlled by capturing and then treating the captured vapor/air.

- All excavated soils when stored prior to treatment should be securely covered with plastic liners and these temporary covers need to be checked and maintained to prevent tear or leaks until the storage pile is moved for treatment. The excavated cells should also be lined, when migration possibilities of contaminated runoff exist, during precipitation events.
- During the main treatment activities as specified above, organic or inorganic vapor emissions should be monitored and appropriate control measures (described in Chapter 13) employed to prevent emissions above the allowable level specified by the regulatory agency (EPA or authorized state).
- When treating soils contaminated with explosive wastes, proper safety measures should be employed and care exercised to prevent any explosion during the treatment process. For conducting safe operations, recommendations provided in the handbook (USEPA, 1993c) may be used, when necessary.
- Periodic system monitoring and evaluation should be performed to prevent leaks or spills.
- When reusing treated soils for restoration of a site, care should be taken to re-create the original soil texture. The treated soil should be verified before replacement. This may require the addition of clays, nutrients, or other materials; some of which can be mixed in from clean soils at the site. During these soil mixing activities, BMPs for pre-treatment should be applied (i.e., cover the areas used for storage, mixing or for processing small batches; minimize work in high temperatures or in high wind, etc.).
- There are four main waste streams generated during soil washing: contaminated solids from the soil washing unit, wastewater, wastewater treatment sludges and residuals, and air emissions (Freeman and Harris, 1995). General BMPs for dealing with these residuals are covered in Chapter 12; additional BMPs are provided below.
  - When collecting moisture or liquids from the treatment process, the contaminated aqueous stream should generally be collected in a tank or a lined/containment system. This should prevent the contaminants from mixing with the normal surface water runoff from the area and the surrounding natural watercourse. The contaminated aqueous stream should be treated or disposed of in accordance with the applicable regulation.
  - An enclosed conveyance system, such as a pipeline or a hose, should be used to move contaminated liquids from the soil washing unit to the containers that will be used to store them.
  - Containers that hold residual liquids should be stored in a place in which they cannot be disturbed or ruptured by large equipment. This may require the construction of a residual management unit separate from the treatment and storage areas.
  - During post-treatment, residuals that are nearly pure listed waste (contaminant) or highly concentrated may need to be managed. These wastes should be dealt with extreme caution to avoid all possible risks of cross-media transfer of contaminants and treated or disposed of in compliance with the applicable state and/or federal regulations.
  - Wastewaters containing hazardous constituents and the high levels of suspended matter can generally be treated with conventional wastewater technologies to acceptable regulatory levels. They should be handled on-site as potentially hazardous wastewaters with appropriate spill prevention contingencies. Air emissions from these units should also be evaluated for appropriate control measures (i.e., closed tanks, covers, etc.) as specified in Chapter 13.

- If solid materials such as granulated carbon filters are used, they should be removed carefully from the emissions system to avoid rupturing them and dissipation of the carbon materials. They should be in airtight containers until they can be recycled or properly disposed of.

5.6 WASTE CHARACTERISTICS THAT MAY INCREASE THE CHANCES OF CROSS-MEDIA CONTAMINATION IN SOIL WASHING TECHNOLOGIES

The effectiveness of soil washing treatment technologies could be compromised, and undue cross-media contamination may be caused under certain conditions. Some soils, especially those that are rich in clays or contain high concentrations of mineralized metals or hydrophobic organics, require very large amounts of additives to achieve acceptable remediation endpoints. In addition, complex mixtures of contaminants in the soil may make it difficult to formulate a single suitable washing fluid that will remove all types of contaminants. In such cases, multiple cleaning fluids may need to be used, and therefore, multiple types of residuals will be generated.

When soil washing is used to remediate contaminated soil, one should consider its potential for generating large amounts of hazardous wastes that must be treated or disposed of (sometimes at a great cost), especially when this technology is used in less than ideal situations. In such cases, even when BMPs are applied, the significant volumes of hazardous wastes that are generated and the reduced efficiency with which the overall system is operating can increase the risk of accident or mismanagement, which can in turn increase the risk of cross-media contamination.

Factors that may limit the applicability and effectiveness of the process include:

- Fine soil particles (e.g., silt, clays) may require the addition of a polymer to remove them from the washing fluid.
- Complex waste mixtures (e.g., metals with organics) make formulating washing fluid difficult.
- High humic content in soil may require pre-treatment.
- The aqueous stream will require treatment.

However, some of these limitations could be overcome with various technology-specific modifications and variations, and some by coupling with other processes (such as further separation of fines, using special solvents, etc.). However, doing so may involve higher costs. The following few characteristics could act as impediments to soil washing treatment process and may result in cross-media transfer of contaminants.

- Soils with high silt and clay content (>50% clay and silt) may be problematic due to the difficulty of removing contamination from very fine particles (Lear, 1996; USEPA, 1993a).
- Soils contaminated with a high concentration of mineralized metals or hydrophobic organics (USEPA, 1993a).
- Complex mixtures of contaminants make it difficult to formulate a suitable washing fluid that will remove all different contaminant types. The cost of formulating such a

fluid would also be prohibitive. Sometimes, a single contaminant/compound could also become strongly bound and difficult to remove (USEPA, 1993a).

- Soil washing treatment is generally not designed for highly explosive material.

REFERENCES

Dutta, S. (2002) *Environmental Treatment Technologies for Hazardous and Medical Wastes, Remedial Scope and Efficacy*. Tata McGraw Hill Publishing Company, New Delhi.

Freeman, H. M., and E. F. Harris. (1995) *Hazardous Waste Remediation: Innovative Treatment Technologies*. Technomic Publishing Co., Inc., Lancaster, PA.

U. S. Environmental Protection Agency (USEPA). (1991a) *Guide for Conducting Treatability Studies Under CERCLA: Soil Washing, Interim Guidance (and Quick Reference Fact Sheet)*. Office of Emergency and Remedial Response. EPA/540/2-91/020A and B, September.

USEPA. (1991b) *Innovative Treatment Technologies-Overview and Guide to Information Sources*. Office of Solid Waste and Emergency Response, Washington, DC. EPA/540/9-91/002, October.

USEPA. (1992a) *A Citizen's Guide to Soil Washing-Technology Fact Sheet*. Office of Solid Waste and Emergency Response, Technology Innovation Office. EPA/542/F-92/003.

USEPA. (1992b) *Seminar Publication-Organic Air Emissions from Waste Management Facilities*. Office of Research and Development, Washington, DC. EPA/625/R-92/003, August.

USEPA. (1993a) *Innovative Site Remediation Technology, Soil Washing/Soil Flushing*. W.C. Anderson, ed. Vol. 3. Office of Solid Waste and Emergency Response. EPA 542/B-93/012.

USEPA. (1993b) *Proposed Best Demonstrated Available Technology (BDAT) Background Document for Hazardous Soil*. Office of Solid Waste, Waste Management Division, Washington, DC, August.

USEPA. (1993c) *Approaches for the Remediation of Federal Facility Sites Contaminated with Explosive or Radioactive Wastes*. Office of Research and Development, Washington, DC. EPA/625/R-93/013, September.

USEPA. (1994a) *BMP Development Workshop Summary-Soil Washing and Thermal Treatment*. Office of Solid Waste, Permits and State Programs Division, Washington, DC, August.

USEPA. (1994b) *Engineering Bulletin-Solvent Extraction*. Office of Research and Development, Cincinnati, OH. EPA/540/S-94/503, April.

USEPA. (1994c) *Remediation Technologies Screening Matrix and Reference Guide*. Federal Remediation Technologies Roundtable, Washington, DC. EPA/542/B-94/013, October.

USEPA. (1997) *Best Management Practices (BMPs) for Soil Treatment Technologies*. Office of Solid Waste, Washington, DC. EPA-530-R-97-007, May.

USEPA. (2002) *Technical Approaches to Characterizing and Cleaning Up Brownfields Sites: Railroad Yards*. EPA/625/R-02/007, July.

Chapter 6

Thermal treatment

The thermal treatment technology is primarily used to remove and/or destroy contaminants that are adsorbed or that adhere to soils, solid media, or water.

6.1 DEFINITION AND SCOPE OF THERMAL TREATMENT

Thermal treatment processes employ indirect or direct heat exchanges to desorb, vaporize, or separate volatile or semi-volatile organics from soils or any solid media while largely avoiding combustion (destruction) of these contaminants in the primary unit. Gases or vapors from the thermal process are treated, destroyed, or condensed for reuse (USEPA, 1997).

A number of different thermal treatment technologies are currently being used for the treatment of soils and the solid media. Considering their similarity in cross-media transfer potentials, the following treatment technologies are listed as a few examples of the thermal treatment:

- Thermal desorption
- Catalytic oxidation
- Thermal bonding
- Molten salt oxidation
- Low-temperature thermal treatment (LT^3) system
- Low-temperature thermal aeration
- Anaerobic thermal processor (ATP)
- Rotary desorbers
- Heated conveyors
- Anaerobic pyrolysis.

Two technologies included in this group – molten salt oxidation and anaerobic pyrolysis – differ slightly from the others but have been included here because they resemble the thermal treatment technologies more closely than any other (AWMA, 1993). However, unlike the other technologies in this group, molten salt oxidation and anaerobic pyrolysis destroy at least some portion of the contaminants present in the soil or solid media (USEPA, 1997). In situ thermal treatment technologies have been used as a source treatment in a number of highly contaminated (Superfund) sites in the US between 2015 and 2017 (USEPA, 2020).

The scope of thermal treatment is not limited to the above-listed technologies. Any treatment technology that meets the key features of thermal treatment should generally be considered as thermal treatment. Incineration technologies are covered separately in Chapter 9.

DOI: 10.1201/9781003004066-6

6.1.1 Key features of thermal treatment

- External application of heat to raise the operating temperature is the unique feature of the thermal treatment.
- Most thermal treatments call for operating temperatures that are significantly above ambient temperature and exceed the boiling point of water, i.e. 212°F (100°C).
- Most thermal treatments are generally designed to remove contaminants from the waste matrix.
- These treatment technologies are generally designed to be non-destructive. However, the high operating temperatures used in some thermal treatment systems will result in localized oxidation or pyrolysis (USEPA, 1994b, 1994c).
- The desorption, vaporization, or separation of different contaminants from the soil matrix varies with the type of contaminants. These variations depend significantly on the type of contaminant and the selected operating temperature.
- The residence time, operating temperature, and the expanse of mixing/agitating the contaminated soil matrix or solid media are generally the prime factors in a thermal treatment process.

6.2 THERMAL TREATMENT TECHNOLOGY DESCRIPTION

Thermal desorption is considered the most general and representative form of thermal treatment. A typical schematic thermal desorption system is shown in Figure 6.1.

In thermal desorption, contaminated material is excavated and delivered to the desorption unit. Excavated material is often stockpiled to provide an adequate feed supply for continuous operation of the treatment facility. Typically, before any treatment, large objects are screened

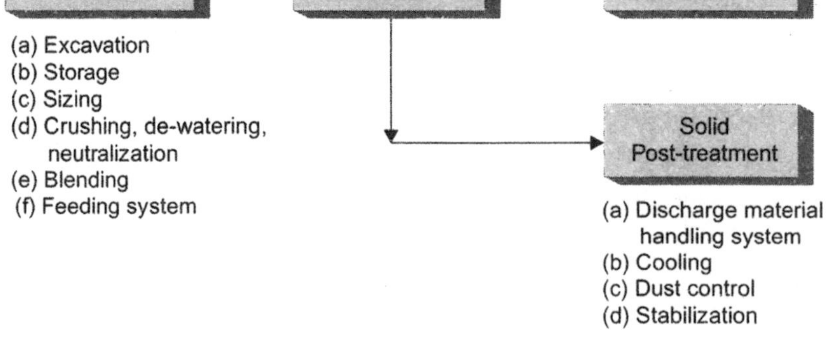

Figure 6.1 A schematic thermal desorption treatment system (USEPA, 1993a).

from the medium and rejected. These rejected materials can sometimes be sized and recycled to the desorber feed. The medium may then be treated to adjust pH and moisture content. It is, thereafter, delivered to the desorber inlet by gravity or conveyed by augers to a feed hopper, from which it is mechanically conveyed to the desorber. In the desorption unit, the contaminated material is heated, and water and contaminants are volatilized. An inert gas may be injected as a sweep stream. Organics in the off-gas may be collected and recovered by condensation and adsorption or burned in an afterburner (USEPA, 1993a).

To increase the effectiveness of thermal desorption technology, extensive pre-processing/pre-treatment of the inlet soil may be conducted. This pre-processing may include removing rocks and debris from the waste matrix, mixing the waste to create a more homogeneous feed, and screening and crushing the waste matrix to achieve a smaller particle size. During pre-processing, air emission monitoring must be conducted and any fugitive emissions controlled (USEPA, 1990).

Thermal desorption systems can create a number of process residual streams during its operation that may need to be managed: treated media; untreated, oversized rejects; condensed contaminants and water; particulate control-system dust; clean off-gas; and spent carbon or other media, if used (USEPA, 1993a).

For further details on the thermal treatment process, readers may find the references (USEPA, 1993b, 1993c, 1992a, 1992b, 1991a, 1991b), cited in this chapter and listed at the end, to be useful.

To provide information on a typical site application of thermal treatment, summary details of a low-temperature thermal desorption (LTTD) and a high-temperature thermal desorption (HTTD) system is furnished below (USEPA, 1994d).

Low-Temperature Thermal Desorption (LTTD).

LTTD is an ex situ process that physically separates volatile organic compounds (VOCs) from the waste matrix. These VOCs or organic gases are then collected or destroyed in a secondary treatment system. The LTTD treatment system consists of several components including feed soil preparation and pre-treatment, treatment of the soil in the desorption unit, solids post-treatment, gas post-treatment, and residuals post-treatment. The most common thermal desorber units include rotary dryers, thermal screws, and belt conveyer systems. Low-temperature desorption usually takes place at a desorber temperature of 300–600°F (149°C–316°C). Higher temperature units are used to remediate less volatile constituents and may be operated at temperatures of 1,200°F (649°C) or so. A typical treatment performance data for LTTD are presented in Table 6.1.

LTTD can be used for most VOCs and may also be used for semi-volatile organic compounds (SVOCs), polychlorinated biphenyls (PCBs), pesticides, and herbicides. Low-temperature systems are not effective with inorganics; higher temperature systems may be used to remediate volatile metals (e.g., mercury).

High-Temperature Thermal Desorption (HTTD).

HTTD is a full-scale technology in which the contaminated media are heated to 320°C–560°C (600 and 1,000°F) to volatilize water and organic contaminants. A carrier gas or vacuum system transports volatilized water and organics to the gas treatment system. HTTD systems involve physical separation processes and are not designed to destroy organics. Bed temperatures and typical residence times will cause selected contaminants to volatilize but not be oxidized. A typical HTTD system is shown in Figure 6.2.

HTTD is frequently used in combination with incineration, solidification/stabilization, or dechlorination, depending on site-specific conditions. This technology has proved that it

Table 6.1 Performance of low-temperature thermal desorption (Dutta, 2002)

Site/vendor	*Contaminants (VOCs/ SVOCs)*	*Performance data (ppm or mg/kg)*			*UTS*[a] *(ppm)*	*Cost*[b]	*Soil volume*	*Remarks*[c]
		Before	*After*	*% Removal*				
Canonie Environmental	Tetrachloroethene	4.9–1,200	<0.025	>99.5	6			
	Toluene	87–3,000	<0.025–0.11	>99.9	10			
	Ethyl benzene	50–440	<0.025-0.14	>99.9	10			
	Xylene	170	<0.16	>99.9	30			
Canonie Environmental	SVOCs	0.7–15	ND-1.0	>93.3	NA			
Letterkenny Army Depot. Roy F. Weston	Benzene	590	0.73	99.9	10		7.5 tons	Data Source[e]
	Trichloroethene	2,680	1.8	99.9	6			
	Tetrachloroethene	1,420	1.4	99.9	6			
	Xylene	27,200	0.55	>99.9	30			
Maine; Smith Environmental Technologies Corp.	Tetrachloroethene	120	ND[d]	>99.9	6	$977/ cubic yards	11,500 cubic yards	liquid waste treatment, storage and disposal site. Data Source[f,g]
	Benzene	2.7	ND	>99.9	10			
	Ethyl benzene	130	ND	>99.9	10			
	Toluene	62	ND	>99.9	10			
	Xylene	840	ND	>99.9	30			
	1,2-Dichloroethylene(o)	300	ND	>99.9	6			
	1,2-Dichlorobenzene	320	ND	>99.9	6			
Washington State, Enviro Klean Systems, Inc.	Benzene	19				$66/ cubic yards	350 cubic yards	Data Source[g]
	Ethyl benzene	71	ND	>99.9	10			
	Toluene	21	ND	>99.9	10			
	Xylene	84	ND	>99.9	30			

[a] UTS – Universal Treatment Standards (non-wastewater).
[b] Current (2020) cost figures are calculated on the basis of the costs listed in the reference(2002) for the Table.
[c] US Environmental Protection Agency requires that hazardous constituents in the soil or solid media must not exceed 10 times the universal treatment standard or UTS. Weblink: https://www.epa.gov/hw/treatment-standards-hazardous-wastes-subject-land-disposal-restrictions#apply .
[d] ND - non-detectable concentration; UST – Underground storage
[e] U.S. EPA, Engineering Bulletin, Thermal Desorption Treatment,@ Office of Research and Development, EPA-540-S-94-501, February 1994.
[f] U.S. EPA, Vendor Information System for Innovative Treatment Technologies (VISITT) - Version 5.0,@ Office of Solid Waste and Emergency Response, August 1996.
[g] Federal Remediation Technologies Roundtable, Remediation Case Studies: Thermal Desorption, Soil Washing, and In-Situ Vitrification,@ EPA-542-R-95-004, March 1995 tank.

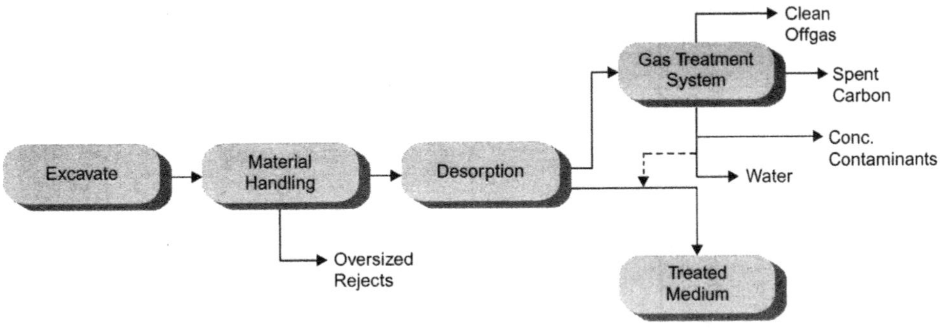

Figure 6.2 A schematic of a typical high-temperature thermal desorption system (Dutta, 2002).

can produce a final contaminant concentration level below 5 mg/kg (ppm) for the target contaminants identified. The target contaminants are SVOCs, PAHs, PCBs, and pesticides. However, HTTD systems have varying degrees of effectiveness against the full spectrum of organic contaminants. VOCs and fuels also may be treated, but treatment may not be very cost-effective. Volatile metals could be removed by HTTD systems. The presence of chlorine can affect the volatilization of some metals, such as lead. The process is applicable for the separation of organics from refinery wastes, coal tar wastes, wood-treating wastes, creosote-contaminated soils, hydrocarbon-contaminated soils, mixed (radioactive and hazardous) wastes, synthetic rubber processing wastes, and paint wastes.

There are a few vendors who are actively implementing this treatment technology. Most of the hardware components for HTTD systems are readily available off the shelf. The time to complete cleanup of a "standard" 18,200-metric ton (20,000-ton) site using HTTD is just over 4 months.

An estimated overall cost is between $187 and $562 per metric ton ($170 and $511 per ton), calculated for the present time (2020), based on the cost provided in the reference (Dutta, 2002).

6.2.1 Advantages and disadvantages of thermal treatment

Some of the technological capabilities and limitations of thermal treatment systems are listed below:

- Thermal treatment is very effective in reducing the volume of the waste while treating the hazardous waste components to an inert level.
- In some applications, it also generates thermal and/or electrical energy from the treatment of the waste.
- It minimizes contaminant migration to soil, air or water when the technology is used with appropriate scrubber systems and proper monitoring of the treated waste before disposal.
- Thermal treatment at intermediate temperatures is widely considered as an effective technology to remove organic contaminants from soil by volatilization and/or destruction.

- Thermal treatment also alters the physical and chemical properties of the soil, and thus affects the leachability of co-contaminants such as heavy metals.
- Thermal technologies in the intermediate temperature range offer an opportunity to destroy simultaneously organic contaminants and to immobilize heavy metals in soil (Zevenbergen et.al., 1997).

Some waste types and operating conditions that may reduce the effectiveness of thermal treatments are provided in detail in Section 6.6. A few limitations of thermal treatment systems are listed below.

- Thermal desorption systems must be closely monitored to ensure that no harmful chemicals, such as dioxins and furans are released into the environment.
- There are chances of explosion for certain contaminants when their concentration exceeds 10–25% of its lower explosive limit (USEPA, 1994a).
- Very high moisture content (>50%) generally makes most thermal treatments cost-intensive (USEPA, 1994c).
- A treatment storage and disposal facility performing thermal treatment of organic waste may be subject to additional analytical requirements depending on the waste management activities that it performs. The permitting authorities may require the facility to comply with the required operating parameters and emission limitations (USEPA, 2015).

6.3 AIR EMISSION AND PUBLIC PARTICIPATION ISSUES INVOLVING THERMAL TREATMENT TECHNOLOGY

As mentioned in Chapter 9, a few common/typical questions and concerns raised by public interest groups and individuals before the implementation of a high-temperature thermal treatment system in the US are furnished below:

- Has air dispersion modeling been performed to demonstrate theoretical compliance with the permitted emissions maximum limitation?
- What criteria were used to maximize public health safety and minimize exposures?
- Commitment to waste stream separation and reduction.
- An oversight committee to continually analyze waste stream components and to evaluate purchasing decisions.
- Any toxicity characteristic leaching procedure (TCLP) test requirements specified in the permit for batches of treated products or residues?
- Is continuous monitoring required?
- Has the facility prepared the package for submittal to regulatory authorities for issuance of permits under the latest effective regulation?
- Has a thorough and independent analysis of options conducted before committing taxpayers funds to a high-temperature thermal system?
- Is open flame or oxidation involved in the thermal system?
- Has a proper, effective, and legitimate lead agency (out of the many state, local, and federal/central Governmental agencies that normally get involved) been identified to coordinate all regulatory issues, compliance, and monitoring, for the proposed system?
- Have all inter-agency and internal memos related to the installation of the proposed system been made accessible to public for information, review and comments on the issue?

- How will the community be kept informed of both routine and unexplained releases from the facility?
- Will soot blowing/grate cleaning be scheduled during daylight hours?
- Has a well-publicized and transparent public opportunity been provided for an extensive technical review of the proposed system?
- Any fugitive dust emission expected at the site from excavation or movement of heavy equipment?

However, a number of best management practices/options (BMPs) are available for mitigating and suppressing fugitive dust emission and other emissions from the use of thermal desorption systems as described later in this chapter. The proper use of such BMPs will also help build community alliance and support in cleanup actions involving thermal treatment systems.

6.4 CROSS-MEDIA TRANSFER POTENTIAL OF THERMAL TREATMENT TECHNOLOGIES

6.4.1 General

The environmental impact associated with all thermal desorbers, aside from process emissions, are attributable to excavation of contaminated solids, management of treated solids, and equipment noise (USEPA, 1993a).

General cross-media transfer potentials during site preparation, pre-treatment, and post-treatment activities are addressed in Chapter 12.

6.4.2 Additional concerns for thermal treatment technologies

- During various thermal treatment operations, SVOC/VOC emissions can occur from leaks in pipes, joints, valves, and uncovered conveyor systems.
- Stack emissions from the collection or destruction of vapors past the thermal desorption, vaporization, or separation treatment unit can release contaminants into the air at levels above the regulatory limit.
- The discharge of scrubber liquor and blowdown can release contaminants to air, water, or soils.
- Waste handling associated with the thermal treatments can contribute significantly to VOC emissions during remediation of soils or solid media.
- Inadequate control and management of baghouse dust containing ash, metals, and/or unoxidized compounds may also cause contaminants to be released into the environment.
- When using any thermal treatment for remediation of explosive wastes, there is the potential risk of possible explosion or detonation of the waste during the treatment process.
- When radioactive/mixed wastes are remediated by thermal treatment, the radionuclides are generally retained or bonded and rendered unleachable for safe disposal of the solid residuals (e.g., ash) in landfills. The likelihood of radioactive emission from the treated mass may still exist.
- Fugitive emissions from fuel sources can sometimes add to the overall emissions of organics from the site.

- In cases where organic wastes are extracted or concentrated rather than destroyed using a thermal treatment technology, the VOC emissions from these wastes can significantly increase the need for control of VOCs from the overall process.
- During remediation of chlorinated organics using the Anaerobic Thermal Processor or any other thermal treatment, there is potential for the emissions of dioxin or dibenzo-furans at low concentrations. Fugitive VOC emissions from the vapor cooling system are also possible (USEPA, 1993a).

6.5 BMPs/OPTIONS FOR CONTROLLING THE CROSS-MEDIA TRANSFER POTENTIAL OF THERMAL TREATMENT TECHNOLOGIES

General BMPs to prevent potential cross-media transfer of contaminants during cleanup activities are addressed in Chapter 12. Also, a proper system design is recommended before implementation of the remedial treatment to avoid cross-media transfer problems during different treatment steps. However, BMP options to control specific cross-media transfer of contaminants for thermal treatments pertaining to the specific treatment activities are given below.

During the thermal treatment process, the following activities are most commonly undertaken:

- Application of heat in a heat exchanger unit;
- Rotational or other mixing action;
- Movement of the contaminated media through a conveyor system;
- Vaporization, desorption, separation, or permanent bonding/solidification of contaminants.

To prevent cross-media transfer of contaminants related to these activities the following BMPs are recommended, where appropriate:

- Fuel storage and fuel handling areas may be added under monitoring and emission control oversight if deemed necessary.
- Routine inspections of pipes, valves, and fittings should be performed where fuel or pressurized liquids are involved.
- During the main treatment activities (specified above), organic or inorganic vapor emissions should be monitored and appropriate emission control measures, as described in Chapter 13 be implemented to prevent emissions above the allowable level specified by the regulatory agency.
- Technology design should take corrosion into account and incorporate corrosion-resistant surfaces for all appropriate pipes, valves, fittings, tanks, and feed systems. It is important that the air pollution control devices are designed for the corrosive nature of the hot gases expected to enter them. Operational plans should include adequate inspection procedures that look specifically for corrosion and wear.
- Operation of thermal desorption systems may create up to six process residual waste streams: treated soil; oversized soil and debris rejects; condensed contaminants and water; spent aqueous and vapor phase activated carbon; and clean off-gas. The following BMPs can be used to control the potential cross-media transfer of residuals:

- Treated medium, debris, and oversized rejects may be suitable for reuse onsite. If not, they should be properly stored or containerized until they can be treated and disposed of.
- The vaporized organic contaminants can be captured by condensation of the off-gas passing through a carbon absorption bed or another treatment system.
- Liquid collection tanks and secondary containment should be incorporated into the operational plans. Subsequent treatment of these concentrated liquids in appropriately regulated units should also be included in the plans.
- Aqueous wastes from scrubber liquors and blowdown could be effectively managed and controlled using various wastewater treatment technologies (Dutta, 2002). Technology operational plans should provide details on proper management and control of these liquids. Further information on treatment technologies are available in references cited at the end of this chapter (USEPA, 1988a, 1987a, 1986a, 1984a)
- Off-gas condensate may contain significant contamination and may require further treatment, such as carbon absorption. If the condensed water is relatively clean, it may be used to suppress the dust from the treated medium.
- Spent granulated carbon should either be returned to the supplier for reactivation or incineration or it should be regenerated on-site.
- System monitoring and evaluation should be performed, as appropriate, to determine possible emissions or migration of contaminants during treatment activities.

6.6 WASTE CHARACTERISTICS THAT MAY INCREASE THE CHANCES OF CROSS-MEDIA CONTAMINATION IN THERMAL TREATMENTS

The following factors may limit the applicability and effectiveness of the thermal treatment process:

- Feed particle size greater than 2 inches can impact applicability or cost at specific sites.
- Dewatering of soils or solid media may be necessary to reduce the amount of energy required to heat the soil.
- Highly abrasive feed can potentially damage the processor unit.
- Clay, silty soils, and soils with high humic content increase the reaction time as a result of binding of contaminants.

In addition, the effectiveness of thermal treatment technologies could be compromised under the following conditions and could cause undue cross-media contamination.

- Solid particles greater than 1.5 inches in size may have to be crushed before treatment to ensure their effectiveness.
- Some desorption systems have produced dioxins and furans through the course of remediating organic contaminants. Desorption systems must be closely monitored to ensure that no harmful chemicals are released to the environment.
- The presence of oil and grease in the soil could reduce the efficiency of thermal desorption systems.

- The contaminant concentration exceeding 10%–25% of its lower explosive limit (USEPA, 1994a).
- Corrosion in containers and equipment due to low (<5 Standard Unit (SU)) or high (>11 SU) pH may pose problems (USEPA, 1994c).
- Particles greater than 2 inches in diameter may have to be treated separately or disposed of due to size limitations for most equipment used for thermal treatment systems (USEPA, 1994c).
- Very high moisture content (>50%) generally makes most thermal treatments highly cost-intensive and might increase the chances for cross-media release of contaminants through the vapor phase (USEPA, 1994c).
- Soils mixed with tars and organic materials, comprising over 10% by volume or weight, may cause handling problems and thus may require the use of a reactor or other equipment to process wastes, which could result in uncontrolled releases of contaminants due to corrosion (USEPA, 1994c).

However, some of these limitations could be overcome with various technology-specific modifications and variations.

REFERENCES

Air & Waste Management Association (AWMA). (1993) *Thermal II Changing Molecular & Physical Status*. AWMA Live Satellite Seminar.

Dutta, S. (2002) *Environmental Treatment Technologies for Hazardous and Medical Wastes, Remedial Scope and Efficacy*. Tata McGraw Hill Publishing Company, New Delhi.

USEPA. (1990) *Handbook on In Situ Treatment of Hazardous Waste-Contaminated Soils*. USEPA/540/2-90/002.

USEPA. (1991a) *Innovative Treatment Technologies-Overview and Guide to Information Sources*. Office of Solid Waste and Emergency Response. EPA/540/9-91/002, October.

USEPA. (1991b) *Engineering Bulletin-Thermal Desorption Treatment*. Office of Research and Development, Cincinnati, OH. EPA/540/2-91/008, May.

USEPA. (1992a) *Guide for Conducting Treatability Studies Under CERCLA: Thermal Desorption Remedy Selection, Interim Guidance (and Quick Reference Fact Sheet)*. Office of Solid Waste and Emergency Response. EPA/540/R-92/074A and B, September.

USEPA. (1992b) *A Citizen's Guide to Thermal Desorption-Technology Fact Sheet*. Office of Solid Waste and Emergency Response, Technology Innovation Office. EPA/542/F-92/006.

USEPA. (1993a) *Innovative Site Remediation Technology, Thermal Desorption*. W. C. Anderson, ed. Vol. 6. Office of Solid Waste and Emergency Response. EPA/542/B-93/011.

USEPA. (1993b) *Approaches for the Remediation of Federal Facility Sites Contaminated with Explosive or Radioactive Wastes*. Office of Research and Development, Washington, DC. EPA/625/R-93/013, September.

USEPA. (1993c) *Proposed Best Demonstrated Available Technology (BDAT) Background Document for Hazardous Soil*. Office of Solid Waste, Waste Management Division, August.

USEPA. (1994a) *BMP Development Workshop Summary for Soil Washing and Thermal Treatment*. Office of Solid Waste, Permits and State Programs Division, August.

USEPA. (1994b) *Innovative Site Remediation Technology, Thermal Destruction*. W. C. Anderson, ed. Vol. 7. Office of Solid Waste and Emergency Response. EPA 542/B-94/003.

USEPA. (1994c) *Engineering Bulletin-Thermal Desorption Treatment*. EPA/540/S-94/501, Office of Research and Development, Cincinnati, OH, February.

USEPA. (1994d) *Remediation Technologies Screening Matrix and Reference Guide*. EPA/542/B-94/013. October.

USEPA. (1997) *Best Management Practices (BMPs) for Soil Treatment Technologies*. EPA-530-R-97-007. Washington, DC, May.

USEPA. (2015) *Waste Analysis Facilities that Generate, Treat, Store, and Dispose of Hazardous Waste – Final*. EPA 530-R-12-001, April.

USEPA. (2020) *Superfund Remedy Report*, 16th Ed. EPA-542-R-20-001, Washington, DC. July.

Zevenbergen, C., and Van Hasselt, H.J. (1997) Immobilisation of heavy metals in contaminated soils by thermal treatment at intermediate temperatures, Studies in Environmental Science Vol. 71, 1997, P. 661-672.

Chapter 7

Vapor extraction

The vapor extraction treatment technology is primarily used to remove and treat contaminants as vapors in environmental remediation.

7.1 DEFINITION AND SCOPE OF VAPOR EXTRACTION TREATMENT

Vapor extraction involves the use of vacuum pumps or blowers to produce a negative pressure gradient, which induces air flow through the waste matrix and causes movement of vapors containing volatile organic compounds (VOCs) towards the extraction wells.

VOCs in the pore spaces of soils or solid media are thereby removed and carried above ground through screened extraction wells. Extracted vapors are treated, as necessary, and discharged, where permissible, to the atmosphere or reinjected to the subsurface (Dutta, 2002; USEPA, 1991d, 1995a, 1997). Vapor extraction is effective at cleaning up contaminant zones/piles containing volatile compounds in homogeneous, permeable soils; with the addition of thermal processes, the technology can be extended to semi-volatile organic compounds (SVOCs) (NRC, 1999).

Vapor extraction treatment technology is predominantly used to remove VOCs. Other treatment technologies used to remove VOCs from soils or solid media are listed in Table 7.1 (USEPA, 1994a). It should be noted that site-specific factors and contaminant characteristics may limit the applicability and effectiveness of any of the technologies and treatments listed in Table 7.1. The technologies listed in the table should always be used in conjunction with other relevant information and references, which may have additional information that could be useful in identifying potentially applicable technologies. Sites where VOCs may be found include burn pits, chemical manufacturing plants or disposal areas, contaminated marine sediments, disposal wells and leach fields, electroplating/metal finishing shops, firefighting training areas, hangars/aircraft maintenance areas, landfills and burial pits, leaking collection and system sanitary lines, leaking storage tanks, radioactive/mixed waste disposal areas, oxidation ponds/lagoons, paint stripping and spray booth areas, pesticide/herbicide mixing areas, solvent degreasing areas, surface impoundments, and vehicle maintenance areas.

DOI: 10.1201/9781003004066-7

Table 7.1 Treatment technologies for commonly encountered VOCs (Dutta, 2002)

Treatment technology	*Development status*	*Applicability for waste types*	*Technology function*
Soil sediment and sludge			
1. In situ biological treatment			
Biodegradation	Full	Better	Destruction
Bioventing	Full	Better	Destruction
2. In situ physical/chemical treatment			
Soil flushing	Full	Better	Extraction
Soil vapor extraction	Full	Better	Extraction
3. In situ physical/chemical treatment			
Thermally enhanced SVE	Full	Average	Extraction
In situ vitrification	Full	Below avg.	Extraction/ destruction
4. Ex situ biological treatment			
Composting	Full	Better	Destruction
Cont. solid phase bio. treat.	Full	Better	Destruction
Landfarming	Full	Better	Destruction
Slurry phase bio. treatment	Full	Better	Destruction
5. Ex situ physical/chemical treatment			
Chemical reduction/oxidation	Full	Average	Destruction
Dehalogenation (BCD)	Full	Average	Destruction
Dehalogenation (glycolate)	Full	Average	Destruction
Soil washing	Full	Average	Extraction
Soil vapor extraction	Full	Better	Extraction
Solvent extraction	Full	Average	Extraction
6. Ex situ thermal treatment			
High temp. thermal desorption	Full	Average	Extraction
Incineration	Full	Average	Destruction
Low temp. thermal desorption	Full	Better	Extraction
Pyrolysis	Full	Below avg.	Destruction
Vitrification	Full	Average	Extraction/ destruction
7. Other treatment			
Excavation and off-site disp.	Full	Average	Extraction/ immobilization
Natural attenuation	Full	Better	Destruction
Groundwater, surface water, and leachate			
8. In situ biological treatment			
Co-metabolic treatment	Pilot	Better	Destruction
Nitrate enhancement	Pilot	Better	Destruction
Oxygen enhance. w/air sparging	Full	Better	Destruction
Oxygen enhance. w/H_2O_2	Full	Better	Destruction
9. In situ physical/chemical treatment			
Air sparging	Full	Better	Extraction

(Continued)

Table 7.1 (Continued) Treatment technologies for commonly encountered VOCs (Dutta, 2002)

Treatment technology	*Development status*	*Applicability for waste types*	*Technology function*
Dual phase extraction	Full	Better	Extraction
Hot water or steam flush/strip	Pilot	Average	Extraction
Passive treatment walls	Full	Better	Destruction
Slurry walls	Full	Average	Immobilization
Vacuum vapor extraction	Full	Better	Extraction
10. Ex situ biological treatment (pumping)			
Bioreactors	Full	Better	Destruction
Air stripping	Full	Better	Extraction
Liquid phase carbon adsorption	Full	Better	Extraction
UV oxidation	Full	Better	Destruction
11. Other treatment			
Natural attenuation	Full	Better	Destruction
12. Air emissions/off-gas			
Biofiltration	Full	Better	Ext./destruction
High energy corona	Pilot	Better	Destruction
Membrane separation	Pilot	Better	Extraction
Oxidation	Full	Better	Destruction
Vapor phase carbon adsorption	Full	Better	Extraction

Typical VOCs [excluding fuels, benzene, toluene, ethyl-benzene, and xylene (BTEX), and other contaminants in gaseous form] encountered at many sites include the following:

- *Halogenated VOCs*

    - Bromodichloromethane
    - Bromoform
    - Bromomethane
    - Carbon tetrachloride
    - Chlorodibromomethane
    - Chloroethane
    - Chloroform
    - Chloromethane
    - Chloropropane
    - Cis-1,2-dichloroethylene
    - Cis-1,3-dichloropropene
    - Dibromomethane
    - 1,1-Dichloroethylene
    - Dichloromethane
    - 1,2-Dichloropropane
    - Ethylene dibromide
    - Fluorotrichloromethane (Freon 11)
    - Hexachloroethane
    - Methylene chloride
    - Monochlorobenzene
    - 1,1,2,2-Tetrachloroethane
    - Tetrachloroethylene (Perchloroethylene) (PCE)
    - 1,2-Trans-dichloroethylene
    - 1,1-Dichloroethane
    - 1,2-Dichloroethane
    - 1,2-Dichloroethene
    - Trichloroethylene (TCE)
    - 1,2,2-trifluoroethane (Freon 113)
    - Trans-1,3-dichloropropene
    - 1,1,1-Trichloroethane
    - 1,1,2-Trichloroethane
    - Vinyl chloride

- *Non-Halogenated VOCs*

• Acetone	• Isobutanol
• Acrolein	• Methanol
• Acrylonitrile	• Methyl ethyl ketone (MEK)
• n-Butyl alcohol	• Methyl isobutyl ketone
• Carbon disulfide	• 4-Methyl-2-pentanone
• Cyclohexanone	• Styrene
• Ethyl acetate	• Tetrahydrofuran
• Ethyl ether	• Vinyl acetate

Different types of vapor extraction treatment technologies are currently being used for treating contaminated media. Considering their similarity in the treatment process, the following technologies are listed as a few examples of vapor extraction treatment:

- Soil vapor extraction (SVE)
- Fracture enhanced vapor extraction
- Thermal enhancements of SVE
- Steam injection
- Hot air injection
- Soil venting
- Bioventing (also in Chapter 8, Bioremediation)
- Air sparging
- Multi-phase or dual-phase extraction
- In situ steam stripping.

The scope of vapor extraction treatment is not limited to the above-listed technologies. Any treatment technology that has similar key features listed under Section 7.1.1 in this chapter should be considered as vapor extraction treatment.

Many of the above-listed technologies are used in conjunction with vapor extraction or groundwater technologies. They may even be used to remediate VOCs in saturated soils or groundwater. Some of these technologies that use thermal processes can also be used to treat SVOCs.

A typical schematic vapor extraction system is shown in Figure 7.1.

7.1.1 Key features of vapor extraction technologies

- With respect to vadose zone soils, vapor extraction relies on the ability to produce an advective air flow field throughout the contaminated soils.
- The vacuum gradient, created during extraction, induces air flow through vadose zone soils to volatilize the contaminants.
- Transfers contaminants from soil to air. Vapor treatment following vapor extraction from the subsurface may be required to minimize the discharge of contaminants into the atmosphere (USEPA, 1995c).
- It generally does not destroy contaminants; extracts contaminant vapors for collection and/or treatment.

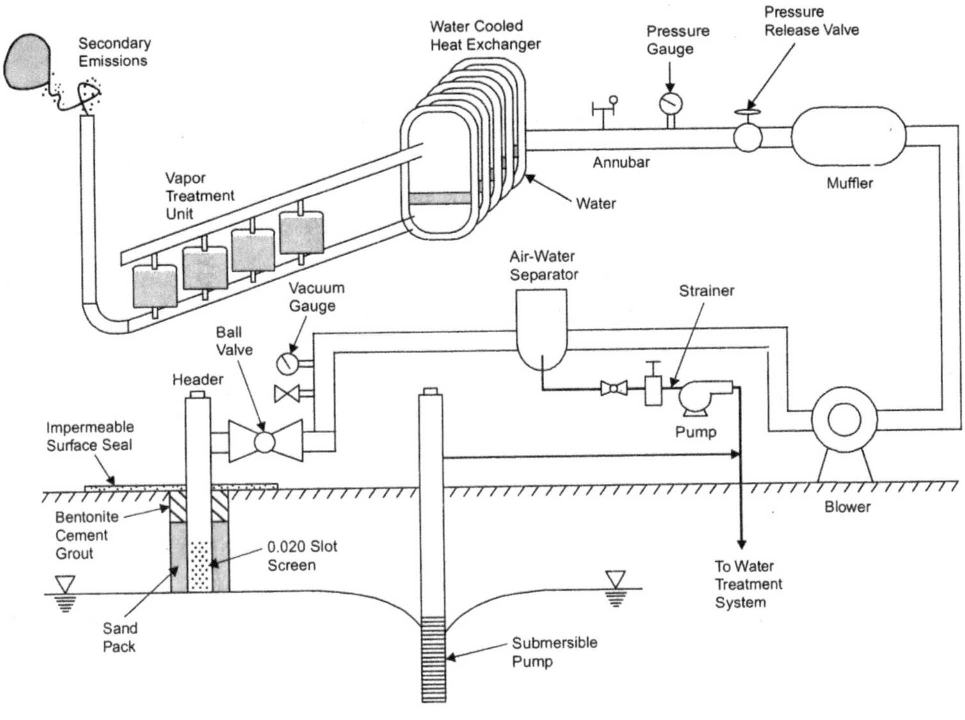

Figure 7.1 A schematic soil vapor extraction system (Dutta, 2002).

- Effectively reduces concentrations of VOCs and certain biodegradable SVOCs. Less volatile contaminants may be removed by bioventing. Heat (hot air or steam) may be applied to increase the volatility of less volatile compounds (USEPA, 1995a).
- Generally, vapor extraction is an in situ technique (excluding above-ground vapor and water treatment). It is primarily designed for use in the vadose zone, although the saturated zone can be dewatered and treated or treated with air sparging combined with vacuum extraction.
- Vapor extraction can be used ex situ for the remediation of aboveground soil piles. For this application, perforated pipes are located within the aboveground soil piles and connected to a blower to draw air through the piles (USEPA, 1995a).

7.2 VAPOR EXTRACTION TECHNOLOGY DESCRIPTION

Vapor extraction processes couple vapor extraction wells with blowers or vacuum pumps to create an air-flow field in soil zones permeable to vapor flow. Contaminants volatilized within the air-flow field are swept into the vapor extraction well and are removed from the soil (Dutta, 2002). Currently, vapor extraction is commonly used as an in situ technology. During

2015–2017, more than 40% of decision documents with source remedies include in situ treatment, such as solidification/stabilization, soil vapor extraction, and in situ thermal treatment (USEPA, 2020).

Vapor treatment systems, such as catalytic or thermal destruction systems, activated carbon adsorbers, or biological gas treatment systems, may be employed to extract and treat the contaminated air stream (USEPA, 1997). Variants of vapor extraction, such as air sparging and thermally enhanced SVE, have been developed to extend the application of vapor extraction technology (USEPA, 1992b, 1995b). Vitrification is one of the thermally enhanced SVE systems. The vitrification process converts contaminated soil to stable glass and crystalline solids. This process produces heat to melt the contaminated soil. The older method uses electrodes and electrical resistance to vitrify materials, while the emerging technique uses plasma arc technology (USEPA, 2020). Further details on in situ vitrification are covered in Chapter 10.

Air sparging extends the application of vapor extraction to water-saturated soils by injecting air under pressure below the water table. Thermally enhanced SVE combines conventional vapor extraction equipment with a means to elevate the subsurface temperature for increasing the volatilization potential of the soil contaminants (USEPA, 1995a).

One advantage of vapor extraction systems is that they generally do not require addition of reagents that must be delivered to and subsequently recovered from the contaminated area (USEPA, 1990). For further details of vapor extraction systems, readers may find it useful to consult the references listed at the end of the chapter (USEPA, 1991b, 1994a, 1994b, 1994c, 1995d, 1995f, 2012, 2020; Dutta, 2002; Johnson et al., 1990; Baehr et al., 1988).

7.3 CROSS-MEDIA TRANSFER POTENTIAL OF VAPOR EXTRACTION TECHNOLOGIES

7.3.1 General

The general cross-media transfer potentials of vapor extraction technologies during site preparation, pre-treatment, and post-treatment activities are addressed in Chapter 12.

7.3.2 Additional concerns pertaining to vapor extraction technologies

- Fugitive vapor emissions through surface soils during operation – especially when injection (air or steam) and fracturing technologies are applied.
- Operation of vapor extraction systems may result in the contaminated soil moisture becoming condensed and entrained with the system and/or cause the uptake of contaminated groundwater.
- Surface water can intrude or channel into the contaminated vadose zone altering the anticipated flow of vapors and groundwater.
- Undesirable migration of subsurface contaminant vapors or liquids into the soil due to improper design or operation. For instance, the improper design of a vapor extraction system can cause unwanted migration of contaminants from source areas to virgin soils, and improper design of an air sparging system can cause unwanted migration

of dissolved phase and vapor phase contaminants to areas outside the remediation zone.

- During vapor extraction system operation, there is potential for VOC emissions through pipes, joints, and valves on the delivery/discharge side of the vacuum pump.
- Inadequate design and operation of emission control equipment may also result in the transfer of contaminants to the natural environment.
- Residual contaminants in the soil after completion of a vapor extraction system operation and the remaining contaminants beyond the effective zone of influence of the extraction wells can act as a continuing source of contamination to groundwater or air.
- Ensuring uniform distribution of subsurface air flow and leaving no "dead" zones within the treatment area are critical factors in the location and design of injection and extraction wells.
- With respect to injection techniques, underground sewer and utility conduits can cause short-circuiting of subsurface air flow and may result in uncontrollable VOC discharges.
- Incomplete cleanup due to low permeability soil strata or units can occur. Subsurface air flow will follow the path of least resistance; therefore, channeling through more permeable soil zones (e.g., around natural gravel lenses, fill materials around buildings or utility lines) can result in a partially or wholly ineffective system. Low permeability areas can have high residual contaminant concentrations while vapor extraction off-gases indicate that the cleanup is complete.

7.4 BEST MANAGEMENT PRACTICES/OPTIONS FOR CONTROLLING THE CROSS-MEDIA TRANSFER POTENTIAL FOR VAPOR EXTRACTION TECHNOLOGIES

General best management practices (BMPs) to prevent the potential migration of contaminants to air, soil, or water, referred as "cross-media transfer" of contaminants, during cleanup activities are addressed in Chapter 12. However, the various BMPs/options to control the specific cross-media transfer of contaminants pertaining to the activities in vapor extraction technologies are specified below:

7.4.1 Additional site preparation and staging BMPs

- Site investigation and operational plans should identify the presence of preferential subsurface air flow pathways and account for all existing underground utilities (e.g., sewers and electrical conduits).
- Design and location of extraction and injection wells are critical to ensure the proper distribution of subsurface air flow within the remediation area.
- Depth to the groundwater table should be identified. Groundwater monitoring wells should be installed to determine the presence of LNAPL free product and its recovery strategies. The SVE system design should consider the presence of free product, and the potential applicability of air sparging for combined vadose zone and saturated zone remediation.
- Extraction wells should not be located near surface water impoundments, underground storm sewers, or other utility conduits.

7.4.2 Additional pre-treatment activities

During the SVE system construction, including the installation of wells and/or horizontal extraction/injection trenches, the following activities are most commonly undertaken:

Trench excavation, well drilling, cuttings storage/treatment, drilling mud control and dewatering, air/dust control for dry-air drilling, and surface protection for contaminated media.

The following BMPs should be used to address these concerns:

- During dry-air drilling of contaminated soils, VOC emissions should be monitored and appropriate emission control measures should be used. For example, the drilling could be configured so that auger cutting can be directed through a large-diameter flexible pipe into bins.
- Details on other emission control measures, which can be applied during drilling activities, are covered in Chapter 13.

7.4.3 Vapor extraction treatment activities

Before SVE system operation, the following activities are generally recommended for better and leak-free operation of the system:

- Determine the total number and orientation of extraction wells.
- Identify the injection wells and determine their configuration.
- Evaluate the permeability and fracturing options for the wells.
- Anchor the wells.
- Install the surface seals when there is potential for uncontrolled emissions.
- Check the system for leaks after completion of construction.
- When VOC emissions or leaks are detected from wellheads, wellhead boxes should be installed using self-sealing neoprene rubber packers to form an effective seal between the casing and upper end of the well screen for wellhead protection.
- Monitor any emissions or discharge from the treatment process and control equipment, such as granulated activated carbon, air stripping, and/or biofiltration should be installed, if necessary. Various technologies to control air emissions are provided in Table 13.3, Chapter 13. Emission control equipment should be operated in conformance with the allowable levels specified by the regulatory authority.
- Conduct local dewatering in areas of open impoundments to control water table and to enhance vadose zone air flow.
- Install pumping systems if necessary, for controlling groundwater table.
- When collecting entrained liquids from vapor extraction technology systems, these liquids should generally be collected in a tank or a lined/contained system. This will prevent the contaminants within the liquids from mixing with the normal surface water runoff from the area, the surrounding natural watercourse, and surrounding soil. These contaminated aqueous streams should be treated or disposed of in accordance with the applicable regulations.

- Conduct monitoring and evaluation of the system performance including general system integrity, changes in groundwater table, and rate of VOC reduction.
- Follow proper safety protocols and precautionary measures to prevent any explosion during the recovery and treatment of any light non-aqueous phase liquids (LNAPLs) generated as a free product. When inlet concentrations of the SVE system exceed 10% of the lower explosion limit (LEL) of the LNAPL, safety measures such as remote operation of the system, temporary shutdown, or intermittent pumping should be considered (NIOSH, 1990; USEPA, 1993).

7.4.4 Post-operation activities

After the extraction wellhead vapor monitoring data reach asymptotic VOC concentrations, soils within the radius of influence of the drawn vacuum may not still be remediated enough to meet the cleanup goal. Thus, post-treatment monitoring will be crucial to ensuring the control of soil vapor generated from passive venting.

The waste streams generated by in situ SVE are vapor, treatment residuals (e.g., spent activated carbon), contaminated groundwater, and soil tailings from drilling the wells. BMPs/options for these waste streams are as follows:

- One option for the vapor stream control/treatment unit is to use a biofiltration process for VOC removal.
- Contaminated granular activated carbon (GAC) should be carefully removed from the adsorption unit and handled, stored, and disposed of or recycled in accordance with manufacturer's recommendations and appropriate state or federal regulations.
- Contaminated groundwater, if extracted along with vapor, may be treated and discharged on-site, if allowed, or collected and treated offsite.
- Proper decommissioning of extraction, injection, and/or monitoring wells should be conducted to prevent any future cross-media transfer of contaminants after the termination of SVE system operation in accordance with state and local regulations.

7.5 WASTE CHARACTERISTICS IN VAPOR EXTRACTION TECHNOLOGIES THAT MAY INCREASE THE CHANCES OF CROSS-MEDIA CONTAMINATION

The effectiveness of vapor extraction technologies could be compromised, and undue cross-media contamination may be caused, under certain conditions identified below:

- Low permeability soils (i.e., $K < 10^{-5}$ cm/s) (USEPA, 1995a).
- The presence of constituents that are not volatile or semi-volatile.
- Sites containing high percentage of silty clay soils and requiring formation fracture or other modifications to meet low soil cleanup levels within a short time period.
- Saturated zone soils may be problematic. Air sparging or multi-phase (dual phase) extraction may be considered to treat saturated zone soils and groundwater.
- Commingled waste may have detrimental effects on the treatment of the vapor (e.g., mercury vapor will significantly affect treatment process used on the vapor emission control).

However, some of these limitations could possibly be overcome by technology-specific modifications, such as fracturing clay layers to increase air permeability and injecting hot air to enhance volatility. However, these modifications would likely increase the design, installation, operation and monitoring costs of the system. Please refer to technology-specific references provided at the end of this chapter for additional information about modifications or variations that can be used to enhance the effectiveness of a vapor extraction technology (USEPA, 1989a, 1989b, 1989c, 1991a, 1991c, 1992a, 1992c, 2012, 2020; Dutta, 2002).

REFERENCES

Baehr, A., G. Hoag, and M. Marley. (1988) Removing Volatile Contaminants from the Unsaturated Zone by Inducing Advective Air-Phase Transport. *Journal of Contaminant Hydrology*, vol. 4, pp. 1–26, June 20.

Dutta, S. (2002) *Environmental Treatment Technologies for Hazardous and Medical Wastes, Remedial Scope and Efficacy*. Tata McGraw Hill Publishing Company, New Delhi.

Johnson, P.C., M.W. Kemblowski, J.D. Colthart, D.L. Buyers, and C.C. Stanley. (1990) *A Practical Approach to the Design, Operation, and Monitoring of In situ Soil Venting Systems*. Shell Development/Shell Oil Company, Spring.

National Institutes of Occupational Safety and Health (NIOSH). (1990) *Pocket Guide to Chemical Hazards*. US Department of Health and Human Services, June.

National Research Council (NRC). (1999) *Groundwater and Soil Cleanup: Improving Management of Persistent Contaminants*. The National Academies Press, Washington, DC. doi:10.17226/9615.

USEPA. (1989a) *Soil Vapor Extraction VOC Control Technology Assessment*. Office of Air Quality Planning and Standards. EPA-450/4-89/017, September.

USEPA. (1989b) *Technology Evaluation Report: SITE Program Demonstration Test -- Terra Vac In Situ Vacuum Extraction System*. Vol. 1. Terra Vac, Groveland, MA. EPA/540/5-89/003a, April.

USEPA. (1989c) *Applications Analysis Report: Terra Vac In Situ Vacuum Extraction System*. Office of Research and Development. EPA/540/A5-89/003, July.

USEPA. (1990) *State of Technology Review-Soil Vapor Extraction Systems*. Hazardous Waste Engineering Research Laboratory. EPA/600/2-89/024, August.

USEPA. (1991a) *Guide for Conducting Treatability Studies Under CERCLA: Soil Vapor Extraction, Interim Guidance (and Quick Reference Fact Sheet)*. Office of Emergency and Remedial Response. EPA/540/2-91/019A and B, September.

USEPA. (1991b) *Engineering Bulletin-Air Stripping of Aqueous Solutions*. Office of Research and Development, Cincinnati, OH. EPA/540/2-91/022, October.

USEPA. (1991c) *Engineering Bulletin-In Situ Soil Vapor Extraction Treatment*. Office of Emergency and Remedial Response, Washington, DC. EPA/540/2-91/006.

USEPA. (1991d) *Reference Handbook-Soil Vapor Extraction Technology*. Office of Research and Development, Washington, DC. EPA/540/2-91/003, February.

USEPA. (1992a) *Conducting Field Tests for Evaluation of Soil Vacuum Extraction Application*. Office of Research and Development, Ada, OK. EPA/540/S-92/004.

USEPA. (1992b) *A Citizen's Guide to Air Sparging. Technology Fact Sheet*. Office of Solid Waste and Emergency Response, Technology Innovation Office. EPA/542/F-92/010.

USEPA. (1992c) *Proceedings of the Symposium on Soil Venting*. Office of Research and Development, Washington, DC, April 29–May 1, 1991, Houston, TX. EPA/600/R-92/174, September.

USEPA. (1993) *Approaches for the Remediation of Federal Facility Sites Contaminated with Explosive or Radioactive Wastes*. Office of Research and Development, Washington, DC. EPA/625/R-93/013, September.

USEPA. (1994a) *Remediation Technologies Screening Matrix and Reference Guide*, Office of Solid Waste and Emergency Response, Washington, DC. EPA/542/B-94/013. October.

USEPA. (1994b) *Soil Vapor Extraction (SVE) Treatment Technology Resource Guide*. Office of Solid Waste and Emergency Response, Technology Innovation Office, Washington, DC. EPA/542/B-94/007.

USEPA. (1994c) *Manual-Alternative Methods for Fluid Delivery and Recovery*. Office of Research and Development, Washington, DC. EPA/625/R 94/003, September.

USEPA. (1995a) *Innovative Site Remediation Technology-Vacuum Vapor Extraction*. Vol. 8. Office of Solid Waste and Emergency Response. EPA 542/B-94/002, April.

USEPA. (1995b) *SITE Demonstration Bulletin, Subsurface Volatilization and Ventilation System, Brown and Root Environmental*. RREL, Cincinnati, OH. EPA/540/MR-94/529, January.

USEPA. (1995c) *SITE Demonstration Bulletin, Unterdruck-Verdempfer-Brunnen Technology (UVB) Vacuum Vaporizing Well*. Roy F. Weston, Inc./IEG Technologies Corporation, RREL, Cincinnati, OH. EPA/540/MR-95/500, January.

USEPA. (1995d) *Review of Mathematical Modeling for Evaluating Soil Vapor Extraction Systems*. Office of Research and Development, Washington, DC. EPA/540/R-95/513, July.

USEPA. (1995f) *How to Evaluate Alternative Cleanup Technologies for Underground Storage Tank Sites*. Office of Solid Waste and Emergency Response. EPA/510/B-95/007, May.

USEPA. (1997) *Best Management Practices (BMPs) for Soil Treatment Technologies*. Office of Solid Waste. Washington, DC. EPA-530-R-97-007, May.

USEPA. (2012) *A Citizen's Guide to Soil Vapor Extraction and Air Sparging*. Office of Solid Waste and Emergency Response, Washington, DC, EPA 542-F-12-018, September.

USEPA. (2020) *Superfund Remedy Report*. 16th Ed. Office of Land and Emergency Management, Washington, DC. EPA-542-R-20-001, July.

Chapter 8

Bioremediation

As naturally occurring processes are increasingly explored and understood, the application of bioremediation treatment is expanding its horizon to uncover many new grounds, sometimes reflecting some of the old and familiar phenomena of nature. Bioremediation or biotreatment primarily represents such natural means to biologically degrade and/or destroy contaminants.

8.1 DEFINITION AND SCOPE OF BIOREMEDIATION

Bioremediation is a treatment technology that uses biodegradation of organic contaminants through stimulation of indigenous microbial populations by providing certain amendments such as adding oxygen or limiting nutrients, or adding exotic microbial species. It uses naturally occurring or externally applied microorganisms to degrade and transform hazardous organic constituents into compounds of reduced toxicity and/or availability. Specific technologies fall into two broad categories: (1) ex situ technologies (e.g., slurry phase, land treatment, solid phase, and composting) and (2) in situ technologies (USEPA, 1991). Active remediation can include the addition of such amendments as nutrients or oxygen while passive remediation utilizes natural attenuation to adequately characterize, model, and monitor the site to evidence natural attenuation and protection of human health and the environment (Dutta, 2002).

Many different types of bioremediation technologies are currently being used for soil treatment. Some of the bioremediation technologies include intrinsic or enhanced bioremediation, which is the focus of this chapter, and can be performed in situ or ex situ under aerobic or anaerobic conditions. During enhanced bioremediation, amendments are typically added to the media to supplement biodegradation processes. Amendments include nutrients (such as nitrogen and phosphorus), electron donors (such as methanol or lactic acid for anaerobic processes), electron acceptors (such as oxygen for aerobic processes, ferric iron or nitrate for anaerobic processes), or microbes for bioaugmentation (USEPA, 2001). The following treatment technologies and processes are listed as a few examples of bioremediation technologies and processes:

DOI: 10.1201/9781003004066-8

- Natural Attenuation
- Aerobic Biodegradation
- Anaerobic Biodegradation
- Biopiles
- Composting
- Land Treatment
- Bioreactors
- Bioscrubbers
- Dehalogenation
- Mycoremediation
- Methanotrophic Process
- Bioremediation of Metals (Changing the Valence)
- Binding of Metals
- Plant Root Uptake (Phytoremediation)
- Fungi Inoculation Process
- Slurry Phase Bioremediation
- Solid Phase Bioremediation
- Bioventing (addressed in Chapter. 7, BMPs for Vapor Extraction)
- Bio Wall for Plume Decontamination (In Situ)

The scope of bioremediation technologies is not limited to the above-listed technologies. Any treatment technology that has similar key features, as described below, should be considered as a bioremediation technology. A typical schematic for solid-phase bioremediation is shown in Figure 8.1.

8.1.1 Key features of bioremediation

- Most bioremediation treatment technologies destroy contaminants in the soil matrix.
- These treatment technologies are generally designed to reduce toxicity either by destroying or by transforming toxic organic compounds to less toxic compounds.
- Indigenous microorganisms, including bacteria and fungi, are most commonly used. In some cases, wastes may be inoculated with specific bacteria or fungi that are known to biodegrade the contaminants of concern. Plants may also be used to enhance biodegradation and stabilize the soil.

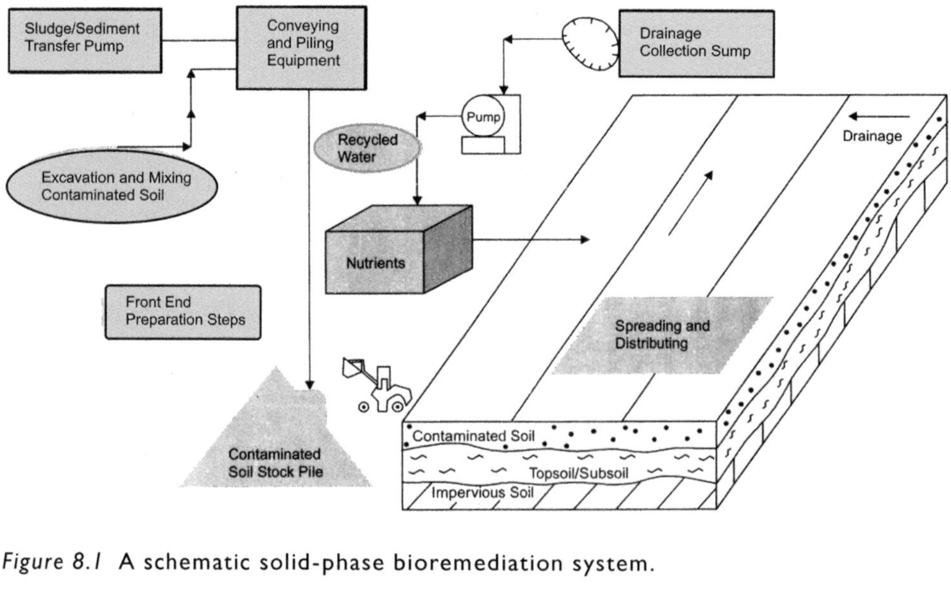

Figure 8.1 A schematic solid-phase bioremediation system.

- The addition of nutrients or electron acceptors (such as hydrogen peroxide or ozone) to enhance the growth and reproduction of indigenous organisms may be required.
- Field applications of bioremediation may involve:
  - Excavation
  - Soil handling
  - Storage of contaminated soil piles
  - Mixing of contaminated soils
  - Aeration of contaminated soils
  - Injection of fluid
  - Extraction of fluid
  - Introduction of nutrients and substrates.

Several of the above field applications need not be viewed separately. Often, multiple applications are used to achieve better results.

8.2 BIOREMEDIATION TECHNOLOGY DESCRIPTION

Bioremediation involves the use of microorganisms to chemically degrade organic contaminants. Aerobic processes use organisms that require oxygen to be able to degrade contaminants. In some cases, additional nutrients such as nitrogen and phosphorous are also needed to encourage the growth of biodegrading organisms. This technology, performed in situ (below ground or in place) or ex situ (above ground), is capable of degrading organic compounds to less toxic materials such as carbon dioxide (CO_2), methane, and water through aerobic or anaerobic processes (USEPA, 2001). A biomass of organisms – which may include entrained constituents of the waste, partially degraded constituents, and intermediate biodegradation products – is formed during the treatment process (USEPA, 1990d).

Bioremediation is being used more often to remediate contaminated media at hazardous waste sites as compared with other remediation technologies. It is often less expensive and more acceptable to the public (USEPA, 2017). The capabilities of Bioremediation technology are often expanding, which, along with its generally lower cost, have made bioremediation an increasingly attractive cleanup technology (Frazar, 2000). Bioremediation was specified as the selected technology for 17% of all superfund site remediation during 2015–2017 involving in situ treatment of soils, sediments, sludges, and groundwater (USEPA, 2020).

The bioremediation process enhances the rate of the natural microbial degradation of contaminants by supplementing these microorganisms with nutrients, carbon sources, or electron donors. This can be performed by using indigenous microorganisms or by adding an enriched culture of microorganisms that have specific characteristics that allow them to degrade the desired contaminant at a quicker rate. Ideally, bioremediation results in the complete mineralization of contaminants to H_2O and CO_2 without the buildup of intermediates. Bioremediation processes can be broadly categorized into two groups: ex situ and in situ. Ex situ bioremediation technologies include bioreactors, biofilters, land farming, and some composting methods. In situ bioremediation technologies include bioventing, biosparging, biostimulation, liquid delivery systems, and some composting methods.

Bioremediation using white-rot fungi to inoculate contaminated media is also being tried at some of the sites. The white-rot fungi was used to treat trinitrotoluene (TNT) and N-methyl-n, 2,4,6-tetranitroaniline with a removal rate of 61% and 53% respectively (USEPA,

2001). This technology can be used in an ex situ or in situ manner. In general, this fungus is used to inoculate a composting process, but it does have other bioremediation applications.

Mycoremediation is another innovative technology using fungi-based uptake and removal of contaminants, which has been explored and used at a few sites. Mycoremediation is a form of bioremediation that uses conditioned native fungi and fungal mycelium applied to surface soils to remove and degrade contaminants. In one of the studies, mycoremediation was used in combination with a bioretention cell (e.g., rain garden), incorporating native vegetation, a soil media mix, and natural microbial assemblages to remove and degrade fecal coliforms and nutrients. The study result indicated that the nutrient removal was increased likely due to the addition of mycorrhizal fungi to the plants in the mycoremediation cell, which might have enhanced plant establishment by increasing the nutrient absorption capacity of root system (Thomas et al., 2009).

The phytoremediation treatment, involving plant uptake/degradation of contaminants, can be used either as an in situ or as an ex situ method. In situ treatments tend to be more attractive to vendors and responsible parties because they generally require less equipment, have lower cost, and cause less disturbance to the pre-existing environment. Further details on phytoremediation treatment are provided in Section 8.3.

Although bioremediation is applied in many different ways, a description of a typical solid-phase bioremediation, composting, bioventing, in situ biodegradation, and a few common bioremediation technologies is provided here.

8.2.1 Solid-phase bioremediation

The solid-phase bioremediation treatment can be conducted in lined land treatment units or in composting piles. A lined land treatment unit consists of a prepared bed reactor with a leachate collection system and irrigation and nutrient delivery systems. The unit may also contain air emission control equipment. Figure 8.1 illustrates treatment by solid-phase bioremediation. The soil is placed on land lined with an impervious layer, such as soil, clay, or a synthetic liner.

8.2.2 Bioventing

Bioventing uses relatively low-flow soil aeration techniques to enhance the biodegradation of soils contaminated with organic contaminants. Although bioventing is predominantly used to treat unsaturated soils, applications involving the remediation of saturated soils and groundwater (augmented by air sparging) are becoming more common. In general, vacuum extraction, air injection, or combination of vacuum extraction and air injection system is employed (Freeman and Harris, 1995). An air pump, one or more air injections or vacuum extraction probes, and emissions monitors at the ground surface level are commonly used.

8.2.3 Landfarming

Landfarming, also known as land treatment or land application, is an above-ground soil remediation technology that reduces concentrations of contaminants, such as petroleum constituents, through biodegradation. This technology usually involves spreading excavated contaminated soils in a thin layer on the ground surface and stimulating aerobic microbial activity within the soils through aeration and/or the addition of minerals, nutrients, and moisture (USEPA, 2017). This is an ex situ treatment process where the contaminated soils are excavated and spread out in a large area for biodegradation.

8.2.4 Bioreactors

Bioremediation may also be conducted in a bioreactor, in which contaminated soil or sludge is slurried with water in a mixing tank or lagoon. Bioreactors function in a manner that is similar to sewage treatment plants. A bioreactor can be designed in many ways, but the majority is a modification of one of two systems. The first system, also known as a trickling filter or fixed media system, allows the aqueous waste stream to trickle over a solid support, such as rocks, that have been colonized extensively by microorganisms. As the liquid waste stream passes over the solids, the microorganisms break down the contaminants. Before the treated waste stream can be discharged, it must be processed through a clarifier so that the number of microorganisms present in the discharge meets regulations. This system can also be referred to as a biofilter, which is also used to treat contaminated gas streams. For this form of treatment to be effective, the contaminant must be volatile.

The second common bioreactor design uses a sealed vessel to mix the contaminants, amendments, and microorganisms. A sealed system allows greater control over factors such as pH and O_2. The waste is pumped into the vessel where it mixes with nutrients and microorganisms. The tank is often aerated, although the process can be kept anaerobic. Multiple bioreactors are often used together to enhance the rate and extent of bioremediation. This system also requires the treatment of the waste stream to remove large numbers of microorganisms before discharge. Bioreactors like this one are often used to treat contaminated solids, which is also known as slurry-phase treatment. In these systems, water is added to the soil and the slurry is continuously mixed. Amendments, such as nutrients, can be added to this system, and treatment can be done under aerobic or anaerobic conditions. The SABRE™ process, which was used in the Bowers Field demonstration in the US, consisted of a slurry-phase bioreactor based on this design to remediate herbicide contaminated soil (USEPA, 1990a). Several variations of these basic designs have been applied for the treatment of pesticides and other hazardous wastes.

For treating contaminated groundwater, variations of either of the two common bioreactor systems are suitable. Conventional pump and treat systems are used to remove the contaminated groundwater from the aquifer. The contaminated water is then treated using either type of bioreactor. The water is then discharged appropriately as required by the regulatory authority. In a liquid delivery system, water is injected into the groundwater through an injection well. This water is amended with nutrients and electron donors prior to injection in order to facilitate increased in situ breakdown of the contaminants by indigenous microorganisms. An air pump is often used to provide oxygen and aerobic conditions. This process produces groundwater, which is extracted and treated if necessary. The extracted groundwater is then either disposed of or amended with nutrients and returned to the contaminated groundwater. The process continues until the extracted groundwater meets cleanup guidelines. Recirculating wells extract the groundwater, amend it, and reinject it without treating it. A schematic for a process where the groundwater is amended and returned for in situ treatment is shown in Figure 8.2.

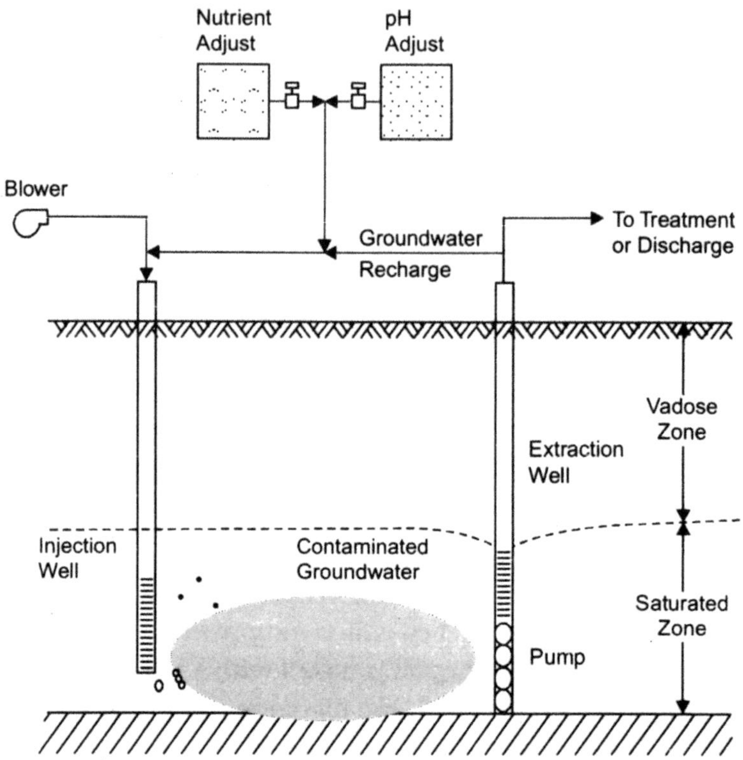

Figure 8.2 Circulation method remediating contaminated groundwater (Singhvi et al., 1994).

8.2.5 Composting

Composting piles is another commonly used solid-phase biological treatment of soil. Figure 8.3 illustrates a typical composting treatment process, including a compost pile, wood chip cover and base, ventilation pipes, and leachate collection system. In composting, the soil is mixed with fertilizer, water, and a bulking agent such as wood chips or sand, and placed in piles from 3 to 6 feet height (USATHAMA, 1988). The bulking agent helps to mix and aerate the material. Oxygen can be supplied to the pile by introducing air (using a blower and series of pipes) below the pile, or by turning/cultivating the pile. The addition of nutrients, such as manure or molasses, to the soil mixture can increase exothermic biodegradation reaction rates and thereby increase the operating temperature of the composting pile. The introduction of air can also help control the temperature of the system. The addition of bulking agents will reduce the concentration of organic compounds in the soil; hence, the concentration of organics in the untreated soil must be sufficiently high to initiate and maintain the composting process. Composting was used successfully at full scale to treat explosives at a site in Oregon, US. The initial concentrations of TNT and 1,3,5-trinitro-1,3,5-triazine (RDX) were 5,250 mg/kg and 1,900 mg/kg, respectively, in the blended soil before the composting treatment. After composting, both of these contaminants were less than 30 mg/kg (USEPA, 2001).

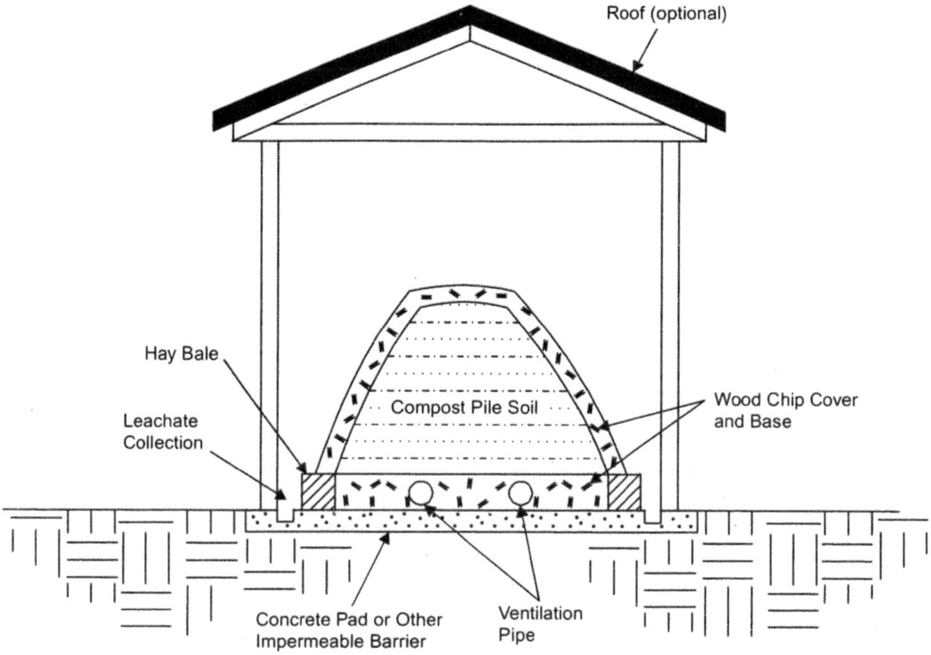

Figure 8.3 Diagram showing solid-phase biological treatment using a composting pile (Dutta, 2002).

8.2.6 In situ biodegradation

The in situ biodegradation process is generally used in conjunction with groundwater pumping and soil flushing systems to circulate nutrients and oxygen through a contaminated aquifer and associated soil (Freeman and Harris, 1995). The process usually involves introducing aerated nutrient-enriched water into the contaminated zone through a series of infiltration galleries with injection wells and recovering the water downgradient. The recovered water can then be treated, if necessary, and reintroduced to the soil on-site. The in situ biodegradation system may also include aboveground treatment and conditioning of the infiltration water with nutrients and an oxygen source or another electron acceptor (USEPA, 1994d).

Bioventing uses relatively low-flow soil aeration techniques to enhance the biodegradation of soils contaminated with organic contaminants. Although bioventing is predominantly used to treat unsaturated soils, applications involving the remediation of saturated soils and groundwater (augmented by air sparging) are becoming more common. In some cases a vacuum extraction, air injection, or a combination of both systems is employed (Freeman and Harris, 1995). An air pump, one or more air injection or vacuum extraction probes, and emissions monitors at the ground surface level are commonly used.

Bioremediation may also be conducted in a bioreactor, in which contaminated soil or sludge is slurried with water in a mixing tank or lagoon. Bioremediation systems require that the contaminated soil or sludge be sufficiently and homogeneously mixed to ensure optimum contact with the seed organisms (USEPA, 1990d)

8.3 PHYTOREMEDIATION

Phytoremediation involves the use of different types of plants to degrade, extract, contain, or immobilize contaminants from soil and water (USEPA, 2000). In recent years, phytoremediation is being used at a number of sites in the US and other countries.

Figure 8.4 shows a schematic of the phytoremediation process. The fundamental basis of this remediation process is root accumulation of contaminants using hydroponic (soil-less) techniques. This is useful for separating inorganic contaminants from water.

Phytoremediation cleanups cause minimal disruption to the site or surrounding community. Initial work may involve grading or tilling of the soil with earth-moving equipment, and backhoes may be needed to plant trees and large shrubs. Any airborne dust during site preparation can be minimized by watering down the soil. Plants used for phytoremediation can make a site more attractive. The use of native plants is encouraged since they are better adapted to the local conditions of the area (USEPA, 2012).

Filtration in the rhizosphere or in rhizomes has been demonstrated at various U.S. Department of Energy (DOE) sites for radionuclides and other heavy metals. Volatilization or transpiration through plants into the atmosphere is another possible mechanism for removing contaminants from the soil or water at a site.

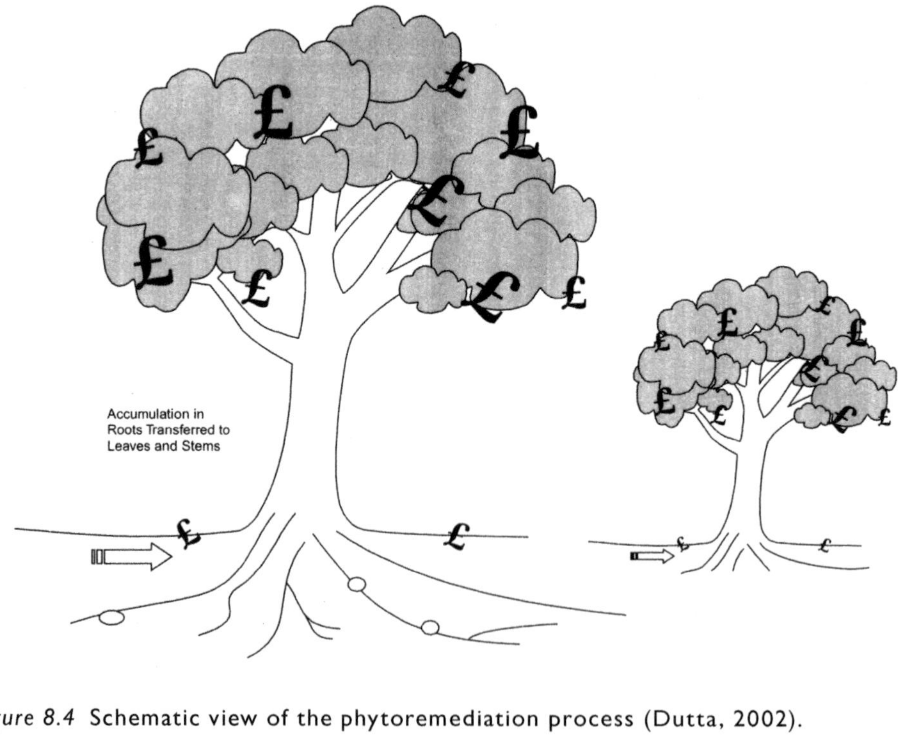

Figure 8.4 Schematic view of the phytoremediation process (Dutta, 2002).

Bioremediation can also be used as a containment technology, where plants either bind the contaminants to the soil, render them nonbioavailable, or immobilize them by removing the means of transport. Physical containment of contaminants by plants can take the form of binding the contaminants within a humic molecule (humification), physical sequestration or segregation of metals occurs in some wetlands, or by root accumulation in plants. Certain trees sequester large concentrations of metals in their roots, and although harvesting and removal are difficult or impractical, the contaminants present a reduced human or environmental risk while they are bound in the roots. Risk reduction may also be achieved by transforming the contaminant into a form that is less toxic, non-toxic or non-hazardous. Mercury (Hg) has also been shown to move through a plant and into the air in a plant that is specifically suited to do so. The thought behind this media switching is that elemental Hg in the air poses less risk than other Hg forms in the soil. However, the technology or the associated risk has not been evaluated yet at great depth.

Phytoremediation applications in the US with specific information on mechanisms, contaminant removed, media treated, and the plants used are summarized in Table 8.1 (Dutta, 2002).

Hydraulic control is another form of containment by the use of phytoremediation treatment. Groundwater contaminant plume control may be achieved by water consumption; using plants to increase the evaporation and transpiration from a site. Some species of plants use a large quantity of water and can extend roots to draw from the saturated zone. Research projects in this area were carried out under the auspices of USEPA at a number of sites, including the Superfund Innovative Technology Evaluation (SITE) program demonstrations at Ogden, Utah, Fort Worth, Texas, the Emergency Response Team (ERT) lead projects at Aberdeen Proving Grounds (Edgewood, MD) and the Edward Sears Properties Site (New Gretna, NJ). Private companies have installed trees as a hydraulic control at many sites. Hydraulic control for groundwater plumes and water balance covers are two technologies applied in the field prior to model development to predict their behavior (Anderson and Walton, 1995).

Vegetative cover (evapotranspiration or water-balance cover) systems are another remediation application utilizing the natural mechanisms of plants for minimizing infiltrating water. Originally proposed in arid and semi-arid regions, vegetative covers are currently being evaluated for all geographic regions. The effectiveness of this technique in all regions and climates needs further evaluation on a site-specific basis. If there is potential for gas generation, a vegetative cover may not be an option. For example, a municipal solid waste landfill can produce landfill gas that may be of concern to human health and the environment. Sites with requirements to collect and control landfill gas may not meet regulatory requirements under the Clean Air Act if a vegetative cover is used.

The use of phytoremediation at hazardous waste sites should be carefully evaluated and studied before the implementation of this technology. As a result of the early information provided by some research and reported by the media, site owners and citizen groups are generally interested in phytoremediation as a clean and cheap technology that may be employed in remediating selected hazardous waste sites. Although current research continues to explore and push the boundaries of phytoremediation applications, there are certain limitations to plant-based remediation systems.

Root contact is a primary limitation on phytoremediation applicability. Remediation with plants requires that the contaminants be in contact with the root zone of the plants. Either the plants must be able to extend roots to the contaminants, or the contaminated media must be brought within a range of the plants. This movement can be accomplished using

Table 8.1 Phytoremediation applications in the US

Mechanism	*Contaminant*	*Media*	*Plant*	*Status*	*Remarks*
Degradation	Atrazine, nitrates	Surface water	Poplar trees	Applied at sites in field scale	For details see listed reference (Schnoor et al., 1995)
Degradation	Landfill leachate	Groundwater	Poplar trees	Applied at sites in field scale	For details see listed reference (Licht, 1990)
Degradation	Trichloroethylene (TCE)	Groundwater	Poplar and cottonwood trees	Field demonstration	For details see listed reference (Rock, 1997)
Degradation	Tri-Nitro-Toluene (TNT)	Wetlands	Various plants	Field demonstration	For details see listed references (Bader, 1996; Carreira, 1996; McCutcheon, 1996)
Degradation	Total petroleum hydrocarbon (TPH)	Soil	Grasses, crops	Field demonstration	For details see listed reference (Banks et al., 1997; Drake, 1997)
Extraction-conc. in shoot	Lead	Soil	Indian mustard	Field demonstration	For details see listed reference (Blaylock et al., 1997)
Extraction-conc. in root	Uranium	Surface water	Sunflower	Field demonstration	For details see listed reference (Dushenkov, 1997)
Extraction	Selenium	Soil, surface water	Various plants	Applied on field scale	For details see listed reference (Bañuelos, 1996; Terry and Bañuelos, 2000)

standard agricultural equipment and practices, such as deep plowing to bring soil from 2 to 3 feet deep to within 8–10 inches of the surface for shallow-rooted crops and grasses, or by irrigating trees and grasses with contaminated groundwater or wastewater. Because these activities can generate fugitive dust and volatile organic compound emissions, potential risks may need to be evaluated. The effective root depth of plants varies according to species and depends upon the soil and climatic condition.

8.4 CROSS-MEDIA TRANSFER POTENTIAL OF BIOREMEDIATION TECHNOLOGIES

8.4.1 General

The general cross-media transfer potentials of bioremediation technologies during site preparation, pre-treatment, and post-treatment activities are addressed in Chapter 12 of this book.

8.4.2 Additional concerns pertaining to bioremediation technologies

- Release of VOCs to air or water can occur during sampling and analysis conducted as part of the implementation process of the bioremediation treatment.
- Some in situ bioremediation processes may generate soil gas emissions in excess of background levels.
- SVOC/VOC emissions may be released from leaks in pipes, joints, valves, and uncovered conveyor systems used in some bioremediation technology operations.
- Nutrients that are applied as a part of the bioremediation treatment, such as landfarming, may be carried through run off to surface water or leach into groundwater if the treatment is improperly designed and/or implemented.
- Very high wind speeds may be of particular concern due to the potential risk of emissions during active composting or other operations where soils may not be covered or enclosed.

8.5 BEST MANAGEMENT OPTIONS FOR CONTROLLING THE CROSS-MEDIA TRANSFER POTENTIAL OF BIOREMEDIATION TECHNOLOGIES

General best management practices/options (BMPs) to prevent the cross-media transfer potential of contaminants during pre- and post-treatment activities have been addressed in Chapter 12. Given below are technology-specific treatment activities and the possible BMP options to control cross-media transfer of contaminants during these activities (USEPA 1997):

During the bioremediation treatment process, the following activities are most commonly undertaken:

- In ex situ bioremediation, excavation, storage, mixing, and other preparatory steps are carried out before feeding the contaminated soil stockpile to the bioreactor or the treatment bed which comprises of lined land treatment units. In the case of composting piles, the soil is mixed with fertilizer, water, and a bulking agent, such as wood chips or sand,

and placed in piles from 3 to 6 feet high (USATHAMA, 1988). In some bioremediation applications, nutrients and substrates are introduced into the treatment bed. Drainage from the treatment bed is generally collected in a sump area and recycled back to the treatment bed.
- In the case of in situ bioremediation, electron acceptors (e.g., oxygen and nitrate), nutrients, and other amendments may be introduced into the soil and groundwater to encourage the growth of indigenous microorganisms capable of degrading the contaminants (USEPA, 1990b). The principal activities surrounding in situ bioremediation involve: (1) boundary determination of the treatment zone in both horizontal extent and depth, (2) injection of nutrients and other amendments, and (3) periodic monitoring of concentration levels of contaminant in the soil or solid media. Bioventing and other in situ biodegradation systems generally use infiltration galleries or injection wells to deliver required amendments to the subsurface.

The following BMPs may be used, when necessary and appropriate, to prevent cross-media transfer of contaminants during the above activities:

- Biotreatments occasionally produce unpleasant odors. Effective odor control measures such as vapor collection systems, or other methods as detailed in Chapter 13, should be used if necessary.
- Surface treatment structures such as biopiles or compost piles should not be constructed and maintained in areas that are likely to encounter high winds. Wind can damage plastic or tarp covers, remove surface materials such as straw or mulch that provide insulation and protection, and carry dust and nutrients away from the treatment area, thereby causing cross-media contamination. In addition, the removal of surface protection increases the likelihood of infiltration of rainfall, which increases the likelihood of the production of leachate.
- For technologies such as composting, landfarming, and other surface treatments, runoff and leachate should be collected and tested regularly to ensure that nutrient levels do not exceed regulatory standards for surface water and groundwater. In particular, nutrients such as phosphate and nitrate may be of concern and may need to be monitored carefully throughout the treatment process.
- Any covers or liners that are used in surface treatment structures, such as biopiles or compost piles, should be periodically examined to ensure that they have not been torn or otherwise damaged. Any damaged covers or liners should be repaired or replaced upon discovery.

8.6 WASTE CHARACTERISTICS THAT MAY INCREASE THE CHANCES OF CROSS-MEDIA CONTAMINATION IN BIOREMEDIATION TECHNOLOGIES

The effectiveness of bioremediation technologies could be compromised and undue cross-media contamination may be caused under certain conditions as given below:

- All bioremediation applications are dependent upon site-specific conditions and the environment or matrix in which the contamination is found. The environment and matrix situation should generally be analyzed before choosing the treatment method. For example, some microbes need specific properties in order to be effective (e.g., high tolerance for acid conditions, saline conditions), and some compounds need to be treated under certain conditions (e.g., aerobically, anaerobically).

- In situ bioremediation applications requiring the circulation of fluids should be avoided in the case of tight clay or heterogeneous subsurface environments where oxygen (or other electron acceptor) transfer limitations exist. This limitation could possibly be overcome by modification of the site or technology, but it would result in a costlier cleanup action.
- In situ bioremediation, generally requires some time to be effective and hence, should not be considered in case of an emergency removal/remedy or where "fast cleanup" is required. However, ex situ treatment, e.g., in a reactor, may be appropriate.
- Concentration of the contaminant above the toxicity tolerance level of the microorganisms (which varies with the bioavailability of the compound) will make biodegradation ineffective.
- Concentrations of contaminants required to sustain microbial population may be above the regulatory limit. Conversely, sometimes, the microbial population cannot survive because of the concentrations of contaminants being too low to sustain it. A treatment train or other alternatives should be considered under these circumstances.
- Technologies to treat some metals in heterogeneous mixtures, such as lead and mercury, should be used very carefully (e.g., iron oxide, hexavalent chromium, and mercury).

Please refer to technology-specific references provided at the end of this chapter for additional information about modifications or variations that can be used to enhance the effectiveness of a bioremediation technology (USEPA, 1990c, 1992a, 1992b, 1992c, 1992d 1995a, 1995b, 1995c).

8.6.1 Additional BMPs for residuals management

In general, there are few residuals to treat with bioremediation technologies as given below:

- Management of treated soils, sediments, and geological material: Often, bioremediation does not bring the concentration of individual hazardous components in soil, subsurface material, or sediment to current concentration-based standards. Depending on the bioavailability of those individual hazardous components to the human population or plants and wildlife, the hazard may or may not be controlled by the treated material.
- Engineered or institutional controls should be in place for the treated soils, as with any technology that is used to treat hazardous wastes, primarily to prevent direct contact with the treated material by the human population or critical environmental receptors. Properly remediated geological material should not release or emit hazardous constituents to soil, air or groundwater in contact with the remediated material.
- If effective engineering or institutional controls are not in place to prevent exposure, a risk analysis of direct exposure should indicate that the level of risk is acceptable.

The exception is the bioreactor, which may generate carbon dioxide, water or other off-gases. Bioreactors should be designed with monitoring equipment to ensure that off-gases do not produce air emissions at levels higher than the state or federal regulatory limit.

REFERENCES

Anderson, T.A., and B.T. Walton. (1995) Comparative Fate of Trichloroethylene in the Root Zone of Plants from a Former Solvent Disposal Site. *Environmental Toxicology and Chemistry*, vol. 14, pp. 2041–2047.

Bader, D.F. (1996) Phytoremediation of Explosives Contaminated Groundwater in Constructed Wetlands. *IBC Conference*, May 8–10, Arlington, VA.

Banks, M.K., A.P. Schwab, and R.S. Govindaraju. (1997) Assessment of Phytoremediation Field Trials. *12th Annual Conference On Hazardous Waste Research*, May 19.

Bañuelos, G.S. (1996) Extended Abstract: Phytoremediation of Se Applied to Soils in Municipal Sewage Sludge. *First International Conference of Contaminants and the Soil Environment*. February 21–28, Adelaide, South Australia.

Blaylock, M.J., D.E. Salt, S. Dushenkov, O. Zakharova, C. Gussman, Y. Kapulnik, B.D. Ensley, and I. Raskin. (1997) Enhanced Accumulation of Pb in Indian Mustard by Soil-Applied Chelating Agents. *Environmental Science & Technology*, vol. 31, pp. 860–865.

Carreira, L.H. (1996) Abstract: Enzymology of Degradation Pathways for TNI International Phytoremediation Conference, May 8–10, Arlington, VA.

Drake, E.N. (1997) Phytoremediation of Aged Petroleum Hydrocarbons in Soil. *IBC Phytoremediation Conference*, June 18–19, 1997, Seattle, WA.

Dushenkov, S., D. Vasudev, Y. Kapulnik, D. Gleba, D. Fleischer, K.C. Ting, and B. Ensley. (1997) Removal of Uranium from Water Using Terrestrial Plants. *Environmental Science & Technology*, vol. 31, no. 12, pp. 3468–3474.

Dutta, S. (2002) *Environmental Treatment Technologies for Hazardous and Medical Wastes, Remedial Scope and Efficacy*. Tata McGraw Hill Publishing Company, New Delhi.

Frazar, C. (2000) *The Bioremediation and Phytoremediation of Pesticide-contaminated Sites*. National Network of Environmental Studies (NNEMS) Fellow, USEPA, Washington, DC, August.

Freeman, H.M., and E.F. Harris. (1995) *Hazardous Waste Remediation: Innovative Treatment Technologies*. Technomic Publishing Co., Inc., Lancaster, PA.

Licht, L.A. (1990) Poplar Tree Buffer Strips Grown in Riparian Zones for Biomass Production and Nonpoint Source Pollution Control. Ph.D. Thesis, University of Iowa, Iowa City, IA.

McCutcheon, S.C. (1996) Phytoremediation of Organic Com-pounds: Science Validation and Field Testing. EPA Workshop Summary on Phytoremediation of Organic Wastes, EPA/600/R-99/107, December 17-19, 1996, Ft. Worth, TX.

Rock, S.A. (1997) Phytoremediation. *Standard Handbook of Hazardous Waste Treatment and Disposal*, H. Freeman (ed.). McGraw Hill.

Schnoor, J.L., L.A. Licht, S.C. McCutcheon, N.L. Wolfe, and L.H. Carreira. (1995) Phytoremediation of Organic and Nutrient Contaminants. *Environmental Science & Technology*, vol. 29, pp. 318A–323A.

Singhvi, R., Koustas, R.N., and M. Mohn. (1994) *Contaminants and Remediation Options at Pesticide Sites*. US EPA. Office of Research and Development, Risk Reduction Engineering Laboratory, Cincinnati, OH. EPA/600/R-94/202.

Terry, N. and G. Bañuelos. (2000) *Phytoremediation of Contaminated Soil and Water*. CRC Press, Taylor & Francis Group, Boca Raton, FL.

Thomas, S.A., L.M. Aston, D.L. Woodruff, and V.I. Cullinan. (2009) *Field Demonstration of Mycoremediation for Removal of Fecal Coliform Bacteria and Nutrients in the Dungeness Watershed*. Battelle, Pacific Northwest Laboratory, Richland, WA, March.

U.S. Army Toxic and Hazardous Materials Agency (USATHAMA) (now U.S. Army Environmental Center (USAEC)). (1988) *Field Demonstration – Composting of Explosives-Contaminated Sediment at the Louisiana Army Ammunition Plant (LAAP)*. Roy F. Weston, West Chester, PA, September.

USEPA. (1990a) *Engineering Bulletin-Slurry Biodegradation*. Office of Research and Development, Cincinnati, OH. EPA/540/2-90/016, September.

USEPA. (1990b) *Engineering Bulletin-In Situ Biodegradation Treatment*. Office of Research and Development. EPA/540/S-94/502, December.
USEPA. (1990c) *Summary of Treatment Technology Effectiveness for Contaminated Soil*. Office of Emergency and Remedial Response, Washington, DC. EPA/540/2-89/053.
USEPA. (1990d) *Handbook on In Situ Treatment of Hazardous Waste - Contaminated Soils*. Office of Research and Development. EPA/540/2-90/002.
USEPA. (1991) *Innovative Treatment Technologies: Overview and Guide to Information Sources*. Office of Solid Waste and Emergency Response. EPA/540/9-91/002, October.
USEPA. (1992a) *Seminar Publication: Organic Air Emissions from Waste Management Facilities*. Office of Air Quality Planning and Standards. EPA/625/R-92/003, August.
USEPA. (1992b) *Engineering Bulletin-Rotating Biological Contactors*. Office of Research and Development, Cincinnati, OH. EPA/540/S-92/007, October.
USEPA. (1992c) *Engineering Bulletin-Slurry Walls*. Office of Research and Development, Cincinnati, OH. EPA/540/S-92/008, October.
USEPA. (1992d) *A Citizen's Guide to Bioventing-Technology Fact Sheet*. Office of Solid Waste and Emergency Response, Technology Innovation Office. EPA/542/F-92/008.
USEPA. (1994d) *Remediation Technologies Screening Matrix and Reference Guide*. EPA/542/B-94/013, October.
USEPA. (1995a) *Innovative Site Remediation Technology-Bioremediation*. Vol. I. Office of Solid Waste and Emergency Response. EPA 542/B-94/006, June.
USEPA. (1995b) *In Situ Remediation Technology Status Report: Treatment Walls*. Office of Solid Waste and Emergency Response. EPA/542/K-94/004.
USEPA. (1995c) *Contaminants and Remedial Options at Solvent-Contaminated Site*. Office of Research and Development. EPA/600/R-94/203.
USEPA. (1997) *Best Management Practices (BMPs). For Soil Treatment Technologies*. Office of Solid Waste, Washington, DC. EPA-530-R-97-007, May.
USEPA. (2000) *Introduction to Phytoremediation*. National Risk Management Research Laboratory, Office of Research and Development, Cincinnati, OH. EPA EPA/600/R-99/107, February.
USEPA. (2001) *Uses of Bioremediation at Superfund Sites*. EPA 542-R-01-019, Washington, DC, September.
USEPA. (2012) *A Citizen's Guide to Phytoremediation*, Washington, DC. EPA 542-F-12-016, September www.epa.gov/superfund/sites.
USEPA. (2017) *How To Evaluate Alternative Cleanup Technologies For Underground Storage Tank Sites*, Washington, DC. EPA 510-B-17-003, October.

Chapter 9

Incineration treatment

The incineration treatment is primarily used to destroy contaminants.

9.1 DEFINITION AND SCOPE OF INCINERATION

Incineration, also known as controlled-flame combustion or calcination, is a remedial technology that destroys organic constituents in soils, debris, or other materials. An incinerator, as defined by the US Environmental Protection Agency, is any enclosed device that uses controlled flame and neither meets the criteria for classification as a boiler, sludge dryer, and carbon regeneration unit, nor is listed as an industrial furnace. Else, it meets the definition of infrared incinerator (electric resistance heater) or plasma arc incinerator (electric arc). Various dictionaries and sources also define incineration as burning, scorching, or carbonization. The American Society of Mechanical Engineers (ASME) defines an incinerator as a device in which wastes are burned (rapidly oxidized) at high temperatures, usually between 1,600°F and 2,500°F, equivalent to 871°C and 1,371°C (Dutta, 2002; USEPA, 1997).

Many different types of incineration technologies are generally used for destroying the contaminants from soils. The incineration treatment is mostly used for destroying or degrading organic contaminants. Because of the stringent emission requirements imposed on incineration technology by regulatory agencies in many countries in the North American and European Continent, the common use of incinerators is on the decline. The following treatment technologies and processes are listed as a few examples of incineration:

- Flame oxidation
- Controlled chamber combustion
- Catalytic oxidation
- Plasma arc and infrared incineration
- Gas or fume incinerator
- Liquid injection incinerators
- Fixed/open hearth incinerators
- Rotary kiln incinerators
- Fluidized bed incinerators

A typical schematic incineration facility is shown in Figure 9.1; a liquid injection incinerator is shown in Figure 9.2; a fixed/sloped hearth incinerator is shown in Figure 9.3, and a multiple hearth and rotary kiln incinerators are shown in Figures 9.4 and 9.5, respectively.

DOI: 10.1201/9781003004066-9

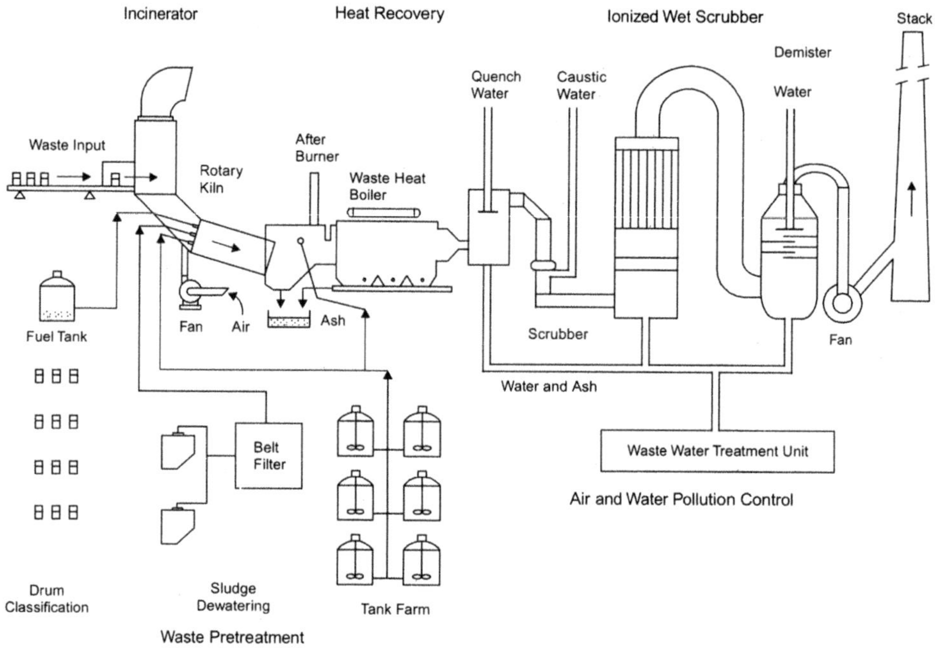

Figure 9.1 A typical schematic incineration facility (Dutta, 2002).

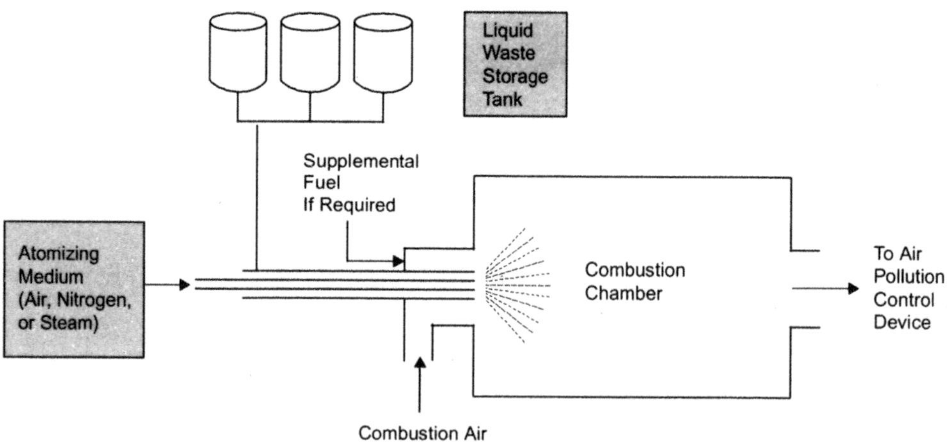

Figure 9.2 A typical liquid injector incinerator (Dutta, 2002).

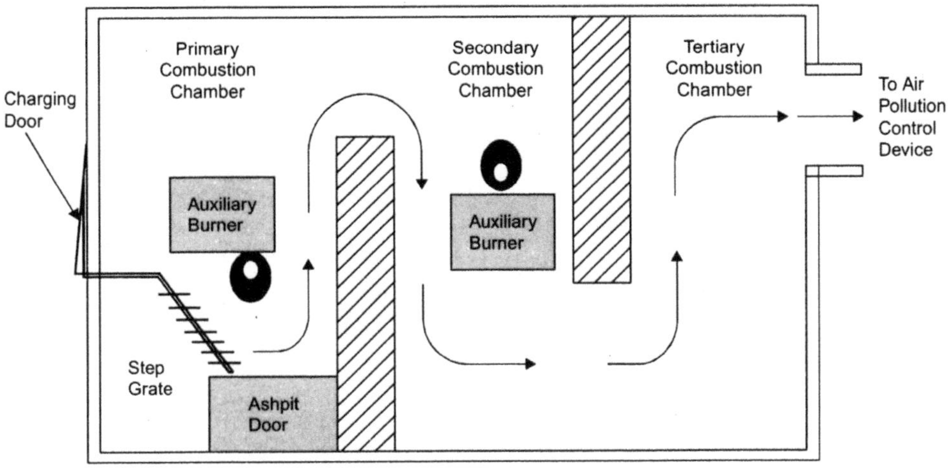

Figure 9.3 A typical fixed/sloped hearth incinerator (Dutta, 2002).

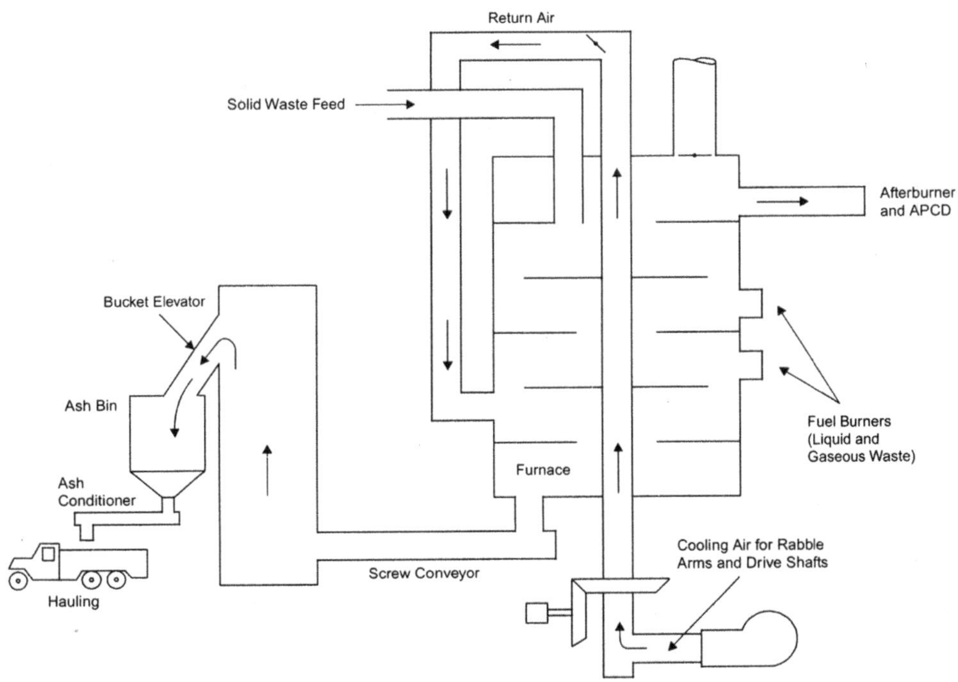

Figure 9.4 Typical multiple hearth incinerator (Dutta, 2002).

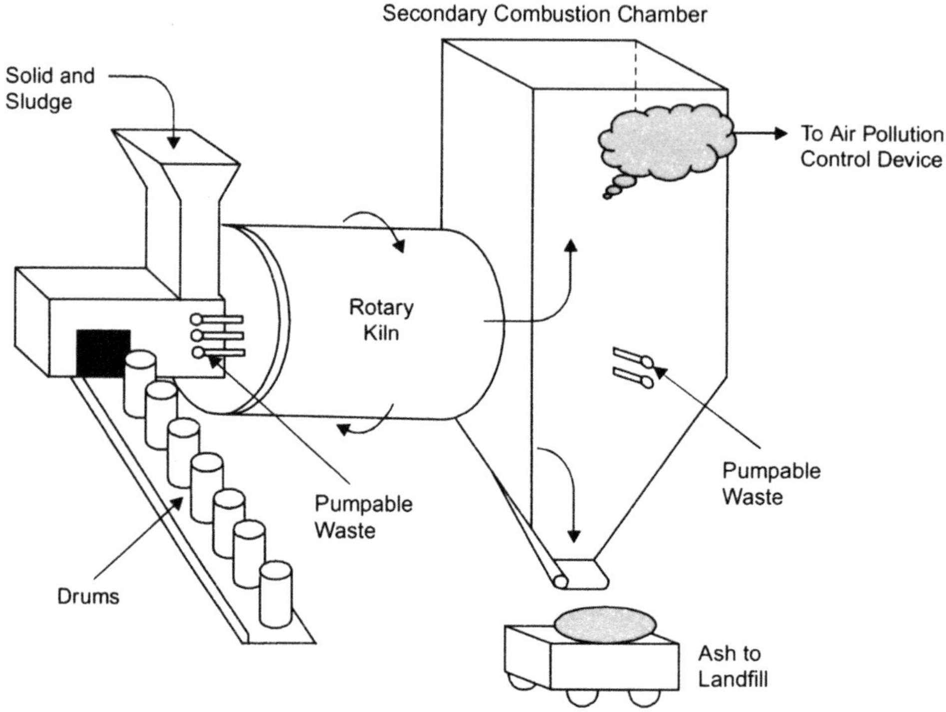

Figure 9.5 Typical rotary kiln incinerator (Dutta, 2002).

9.1.1 Key features of incineration (Dutta, 2002)

- An ex situ treatment.
- Generally involves flame oxidation/burning.
- Produce stack gas emissions.

- Incineration technology generally destroys contaminants in the soil matrix at elevated temperatures.
- Reduces volumes of toxic compounds in the soils.
- Reduces toxicity of organics.
- Does not destroy most inorganic waste (e.g. metals).
- The prime operating factors are the residence time, operating temperature, and the expanse of mixing/agitating the contaminated soil matrix or the solid media (turbulence).

9.2 INCINERATION TECHNOLOGY DESCRIPTION

The incineration process may be viewed as consisting of four steps: (1) preparation of the feed materials for placement in the incinerator, (2) incineration or combustion of the material in a combustion chamber, (3) cleaning of the resultant air stream by air pollution control devices that are suitable for the application at hand, and (4) disposal of the residues from the application of the process (including ash and air pollution control system residues) (USEPA, 1989, 1994).

The initial preparation process involves excavating or pumping the waste product into containers. The pre-processing of feed materials may include screening and mixing as well as crushing to provide a consistent particle size and homogeneity more suitable for the treatment. Although extensive pre-processing will appear to increase capital and O&M costs, these costs are offset by greater levels of efficiency (and lower downtime) with which the system will subsequently operate (USEPA, 2012).

Incineration can destroy a wide range of highly contaminated wastes and greatly reduce the amount of material that must be disposed of in a landfill. For small contaminated areas, excavation and transport to an offsite incinerator can be a quick cleanup approach. Especially, when a site must be cleaned up quickly to prevent immediate harm to people or the environment, a faster cleanup may be important (USEPA, 2012). Incinerators are commonly used to destroy hospital, medical, and infectious waste.

9.3 AIR EMISSION AND PUBLIC PERCEPTION ISSUES INVOLVING INCINERATION TREATMENT TECHNOLOGY

The hospital, medical, and infectious waste incinerator (HMIWI) source category is divided into the following three subcategories based on waste burning capacity:

1. Small (less than 200 lb/h or 91 kg/h),
2. Medium (200–500 lb/h or 91–227 kg/h) and
3. Large (greater than 500 lb/h or 227 kg/h).

Waste burning capacity of an incinerator is determined either by the maximum design capacity or by the maximum charge rate established during the performance test. This generally implies that a default classification based on burning capacity is required. The classification may only be changed after an incinerator is in service, and a maximum charge rate can be established based on the operation. Thus, any concerns raised by the public before installation or commissioning of an incinerator should be given valuable consideration.

Some general questions and concerns expressed by public interest groups and individuals before installation of new incinerators in the US are furnished below (Honican et al., 1998):

- Has air dispersion modeling been performed to demonstrate theoretical compliance with the permit emissions maximum limitation?
- What criteria were used to maximize public health safety and minimize exposures?
- Site analysis had no decision makers who gave appropriate weight to visual aesthetics and possible future development.
- Commitment to waste stream separation and reduction.
- Need an oversight committee to continually analyze waste stream components and to evaluate purchasing decisions.
- Any toxicity characteristic leaching procedure (TCLP) test requirements specified in the permit for batches of ash?
- Is continuous monitoring required?
- Is there any monitoring requirement for radioactive emissions from both medical and pathological waste streams?
- When burning a mixed bag of heterogeneous wastes, is there any radiation monitoring of the effluent streams originating from the incinerator (solid-, liquid-, and gas-phase)?
- Has a thorough and independent analysis of options been conducted before committing taxpayer's funds to an incineration system?
- Has the facility prepared the package for submittal to regulatory authorities for issuance of permits under the latest effective regulation?
- Has a proper, effective, and a legitimate lead agency (out of the many state, local, and federal/central governmental agencies that normally get involved) been identified to coordinate all regulatory issues, compliance, and monitoring, for the proposed incinerator?
- Have all the inter-agency and internal memos related to the installation of the proposed incinerator been made accessible to the public for information, review, and sagacious comments on the issue?
- Can the authorities demonstrate evidence that the proposed incinerator can achieve a 99% efficiency for control of Cl, Hf, and HBr, and other contaminants?
- When it is impossible to know the extent of compounds formulated during the incineration process, and when the mix changes every second, how can the best available control technology for toxics (T-BACT) be expected to control all toxic air pollutants?
- How will the community be kept informed of both routine and unexplained releases from the facility?
- Has the requirement of the maximum available control technology (MACT) floor of a combined dry/wet scrubber for a large hospital/medical/infectious waste incinerator (HMIWI) been met?
- Will soot blowing/grate cleaning be scheduled during daylight hours?
- Has a well-publicized and transparent public opportunity been provided for an extensive technical review of the proposed incinerator facility?

9.3.1 Example project summary (State of Washington)

An example of a typical determination/evaluation by a State Regulatory Agency (Washington Department of Ecology) in the US concerning installation/upgrading of a new incinerator, the relevant permitting requirements, operation and maintenance (O&M), monitoring and recordkeeping requirements are provided in detail for a project below. This should provide an in-depth look into the incineration treatment, its regulatory requirements, operating conditions, construction, safety, emissions, and other technical issues (Dutta, 2002).

State of Washington project summary (example):

Construction and operation of an incinerator to burn infectious medical waste, low-level radioactive waste (LLRW), and pathological wastes.

Facility/representative:

Mr. Gene Patterson, Public Health/Air and Water Quality Manager Washington State University (WSU) Environmental Health and Safety P.O. Box 641172 Pullman, Washington 99164-1172

Location:

New Incinerator Building, Dairy Road, Washington State University (WSU) Campus Pullman a, Washington, Whitman County

Findings of fact:

Laws and regulations

The proposed project by WSU, referred herein to as the permittee, shall comply with all requirements as specified in Revised Code of Washington (RCW), Washington Clean Air Act, Solid Waste Management - Reduction and Recycling, Washington Administrative Code (WAC), General Regulations for Air Pollution Sources, and Controls for New Sources of Toxic Air Pollutants. Specifically, this proposal qualifies as a source of air contaminants as allowed under (Washington State Department of Ecology, 2020):

- WAC 173-400-113, September 13, 1996,
- WAC 173-460-040, January 14, 1994, and
- RCW 70.94.152 – amendment required by RCW 70.94.161 as of August 25, 2020.

Emissions estimate

The following are the estimated process emissions of criteria and toxic pollutants from the project. These estimates are based on the application of Best Available Control Technology (BACT) and Best Available Control Technology for Toxics (T-BACT) as delineated below and as reflected in the APPROVAL CONDITIONS emission limits. Some estimates (as noted) are based on AP-42 emission factors for medical waste incineration. Table 9.1 provides the estimated process emissions in pounds/year (lbs/yr) and the equivalent kilograms/year (kg/yr):

Estimated fugitive emissions, following the application of BACT and T-BACT as delineated below, are as follows:

BACT

WAC 173-400-113(2) requires that BACT be employed for all pollutants. The following technologies and procedures are determined to be BACT:

Table 9.1 Estimated process emissions of pollutants from state of Washington project

Process emission pollutants	*lbs/yr*	*Kg/yr*
Particulate matter (PM)	1,197	544
Carbon monoxide (CO)	506	230
Nitrogen oxides (NO_x)	5,192	2,360
Sulfur dioxide (SO_2)	578	263
Total organic compounds (TOC)	81[a]	37
All toxic air pollutants (TAPs)		
(TAPs) (excluding NO)	201	91.36
Selected toxic air pollutants (TAPs)		
Hydrogen chloride (HCl)	165	75
Mercury (Hg)	9.4	4.27
Cadmium (Cd)	1.2	0.55
(assumes ≤ 2,800 h	20.3	9.23
of operation per year) lead (Pb)		
Dioxins/furans		
Toxic equivalents (TEQ tot)	0.00004[b]	0.000018
*Dioxins/furans, total	0.0039[a]	0.00177
Dioxins/furans (TEQ)	0.0000037[a,b]	(0.0000017)
Chromium (Cr)	0.5964[a]	0.271
(assumes all Cr^{+6})		
Nickel (Ni)	1.47[a]	0.66800
Chlorine (Cl)	0.608[a]	0.276
(at 99% control efficiency)		
Hydrogen fluoride (HF)	0.863[a]	0.392
(at 99% control efficiency)		
Hydrogen Bromide (HBr)	0.251[a]	0.114
(at 99% control efficiency)		
Fugitive pollutants		
(pounds per year) PM	0	0

a Based on AP-42 medical waste emission factors and all waste being medical.
b Expressed in toxic equivalents (TEQ) of 2,3,7,8-tetrachlorinated dibenzo-p-dioxin, per Table 2 of USEPA's regulation 40 CFR 60, Subpart Ec.

Incinerator

The permittee proposed the use of a spray tower, a venturi scrubber, and mist eliminator to control air pollution emissions from the incinerator. These technologies, along with the following, are determined to be BACT for criteria pollutants:

PM: Use of good combustion design, proper waste sizing and loading, proper combustion control, and effective maintenance to control PM emissions.
CO: Use of good combustion design, proper waste sizing and loading, proper combustion control, and effective maintenance to control CO emissions.
NO_x: Use of natural gas as pre-heating and auxiliary fuel for the incinerator.
SO_2: Use of natural gas as pre-heating and auxiliary fuel.
TOC: Use of good combustion design, proper waste sizing and loading, proper combustion control, and effective maintenance to control VOC emissions.

Opacity: Use of good combustion design, proper waste sizing and loading, proper combustion control, and effective maintenance to limit visible stack emissions.

Fugitive emissions – on-site roads.

PM: BACT is determined to be either the pavement of roads, chemical treatment of roads with approved binders, or on-site water application to roads as required.

Fugitive emissions – incinerator ash handling

PM: BACT for the handling of incinerator ash handling is determined to be only transferring ash from the incinerator to sealable metal containers within the enclosed incinerator building followed by disposal at an approved site.

Cooling tower

PM: BACT for the cooling tower is determined to be the use of a high-efficiency drift eliminator (0.005% drift as proposed).

T-BACT

This project is required to use Best Available Control Technology for Toxics (T-BACT) to control TAP emissions. The BACT determination above constitutes T-BACT for all toxic air pollutants. Specifically, the use of treatment chemicals that are not toxic air pollutants constitutes T-BACT for the cooling tower.

Additional findings

- The provisions of Certification of Operators of Solid Waste Incinerator and Landfill Facilities apply.
- The provisions of solid waste incinerator facilities do not apply since the incinerator is not designed to burn 12 or more tons per day.
- The provisions of ambient air quality standards and emission limits for radionuclides apply since this project proposes to burn up to 200 pounds per hour of low-level radioactive waste.
- This project shall utilize the best available radionuclide control technology (BARCT). This is demonstrated by the permittee having a 'radioactive air emissions license' issued by the Washington Department of Health.
- The owner/operator of this incinerator facility must comply with the requirements for an Environmental Impact Statement (EIS).
- Standards of performance are applicable for hospital/medical/infectious waste incinerators, which commenced construction after June 20, 1996. In addition, US EPA regulation 40 CFR 60, Subpart A {General Provisions} applies.

Approval conditions

Waste throughput and fuel consumption limit

Waste throughputs and annual hours of operation shall be limited to the amounts given below:

- Infectious medical waste or low-level radioactive waste: 200 pounds per hour (lbs/h) or 91 kg/h and the total medical and radioactive waste: 180,855 pounds (82,207 kg) per calendar year.
- Pathological waste: 800 pounds (364 kg) per hour and 977,168 pounds (444,167 kg) per calendar year
- Hours of incinerator operation: 2,800 h per calendar year.
- The waste throughput and fuel consumption caps shall not be exceeded until a revised Notice of Construction (NOC) application is submitted to, and a revised approval order issued by the Washington State Department of Ecology (Ecology).

- Any planned modification to the plant or operating procedures shall be reported to Ecology.

General testing requirements

- Within 60 days of achieving the maximum production rate at which the facility will be operated, but no later than 180 days after initial startup, the permittee shall contract the conducting of tests specified below to an independent testing firm and provide a written report of such tests.
- A test plan shall be submitted for approval by Ecology, at least 30 days before any source testing.
- WSU shall provide testable emission points, sampling ports, safe sampling platforms, safe access to sampling platforms, and utilities for sampling and testing.
- During initial and any periodic performance testing, the incinerator shall:
  - o Operate at 200 pounds per hour of medical waste and
  - o During a separate set of tests, operate at a 720–800 pounds (327.2–363.4 kg) per hour of pathological waste.
  - o Conduct testing of both medical and pathological waste streams that consist of three separate runs each with medical and pathological waste.

The tests specified below shall be conducted separately for both medical and pathological wastes:

- Unless alternate or equivalent tests are requested in writing by the permittee and approved of by Ecology or
- unless specifically noted below under Emission Limits & Associated Testing,
- The use of the bypass stack during any performance test shall invalidate that performance test.

Emission limits & associated testing

The following types of emissions shall be tested for, assessed, or monitored and then compared to the corresponding compliance limits listed below:

- Opacity: visible emissions for any emission unit or fugitive source shall be no more than 10%, averaged over 6 minutes.

Associated monitoring conditions:

- A continuous emission monitoring system (CEMS) for measuring opacity from the incinerator exhaust stack to the atmosphere shall be used to determine ongoing compliance.
- Opacity readings as measured according to the US EPA regulation 40 CFR, Part 60, Appendix A, Method 9, may also be conducted during site visits by Ecology certified opacity readers and be used to determine ongoing compliance.

Particulate Matter (PM):

Emissions shall not exceed 0.03 grains per dry standard cubic foot.

Note: "Standard" refers to a temperature of 68°F (20°C) and a pressure of 1 atmosphere.

Associated testing conditions:

- Initial performance testing for the PM emission limit at the exhaust stack to the atmosphere shall be conducted according to 40 CFR, Part 60, Appendix A, Method 5 (front half) and 40 CFR, Part 51, Appendix M, Method 202 (back half). Such testing shall be conducted no later than 90 days after the initial startup of the incinerator.
- Periodic performance testing for PM, according to 40 CFR, Part 60, Appendix A, Method 5 (front half) and 40 CFR, Part 51, Appendix M, Method 202 (back half), shall be conducted once each 12 months for at least the first 3 years. Thereafter:

- If all three performance tests over a 3-year period indicate compliance with the PM emission limit, the owner or operator may forego a performance test for that pollutant for the subsequent 2 years.
- At a minimum, a performance test for PM shall be conducted every third year (no more than 36 months following the previous performance test).
- If a performance test conducted over 3 years indicates compliance with the emission limit for PM, the owner or operator may forego a performance test for that pollutant for an additional 2 years.
- If any performance test indicates noncompliance with the respective emission limit, a performance test for that pollutant shall be conducted annually until all annual performance tests over a 3-year period indicate compliance with the emission limit.

Carbon monoxide (CO):

Emissions shall not exceed 40 parts per million by volume (dry, 7% O_2).

Associated testing conditions:

- Initial performance testing for the CO emission limit at the exhaust stack to the atmosphere shall be conducted according to 40 CFR, Part 60, Appendix A, Method 10. Such testing shall be conducted within the same time frame as the initial PM compliance testing.
- Periodic performance testing for CO, according to 40 CFR, Part 60, Appendix A, Method 10, shall follow the schedule for PM performance testing.

Dioxins/furans:

Emissions shall not exceed 1.0 grains per billion dry standard cubic feet for total dioxins/furans, based on the toxic equivalency factors published in Table 2 of 40 CFR 60, Subpart Ec.

Associated testing conditions:

- Testing for and compliance with the dioxins/furans emission limit shall only be required when combusting medical waste.
- Initial performance testing for the dioxins/furans emission limit at the incinerator exhaust stack shall be conducted according to 40 CFR, Part 60, Appendix A, Method 23. The minimum sample time shall be four hours per test run. Such testing shall be conducted within the same time frame as the initial PM compliance testing.

Hydrochloric Acid (HCl):

Emissions shall not exceed 10 parts per million by volume (dry, 7% O_2).

Associated testing conditions:

- Testing for and compliance with the HCl emission limit shall only be required when combusting medical waste.
- Initial performance testing for the HCl emission limit at the incinerator exhaust stack shall be conducted according to 40 CFR, Part 60, Appendix A, Method 26. Such testing shall be conducted within the same time frame as the initial PM compliance testing.
- Periodic performance testing for HCl, according to 40 CFR, Part 60, Appendix A, Method 26, shall follow the schedule for PM performance testing.

Sulfur dioxide (SO_2):

Emissions shall not exceed 20 parts per million by volume (dry, 7% O_2).

Associated testing conditions:

- Testing for and compliance with the SO_2 emission limit shall only be required when combusting medical waste.

- Initial performance testing for the SO_2 emission limit at the turbine exhaust shall be conducted according to 40 CFR, Part 60, Appendix A, Method 6C. Such testing shall be conducted within the same time frame as the initial PM compliance testing.
- Additional performance testing for SO_2, according to 40 CFR, Part 60, Appendix A, Method 6C, may be required in the future upon written notification by Ecology.

Nitrous oxides (NOx):

Emissions shall not exceed 250 parts per million by volume (dry, 7% O_2).

Associated testing conditions:

- Testing for and compliance with the NOx emission limit shall only be required when combusting medical waste.
- Initial performance testing for the NOx emission limit at the incinerator exhaust stack shall be conducted according to 40 CFR, Part 60, Appendix A, Method 7E. Such testing shall be conducted within the same time frame as the initial PM compliance testing.
- Periodic performance testing for NOx, according to 40 CFR, Part 60, Appendix A, Method 7E, may be required in the future upon written notification by Ecology.

Lead (Pb):

Shall not exceed 0.52 grains per thousand dry standard cubic feet.

Associated testing conditions:

- Testing for and compliance with the Pb emission limit shall only be required when combusting medical waste.
- Initial performance testing for the Pb emission limit at the incinerator exhaust stack shall be conducted according to 40 CFR, Part 60, Appendix A, Method 29. Such testing shall be conducted within the same time frame as the initial PM compliance testing.
- Periodic performance testing for Pb, according to 40 CFR, Part 60, Appendix A, Method 29, may be required in the future upon written notification by Ecology.

Cadmium (Cd):

Shall not exceed 0.031 grains per thousand dry standard cubic feet.

Associated testing conditions:

- Testing for and compliance with the Cd emission limit shall only be required when combusting medical waste.
- Initial performance testing for the Cd emission limit at the incinerator exhaust stack shall be conducted according to 40 CFR, Part 60, Appendix A, Method 29. Such testing shall be conducted within the same time frame as the initial PM compliance testing.
- Periodic performance testing for Cd, according to 40 CFR, Part 60, Appendix A, Method 29, may be required in the future upon written notification by Ecology.

Mercury (Hg):

Shall not exceed 0.24 grains per thousand dry standard cubic feet.

Associated testing conditions:

- Testing for and compliance with the Hg emission limit shall only be required when combusting medical waste.
- Initial performance testing for the Hg emission limit at the incinerator exhaust stack shall be conducted according to 40 CFR, Part 60, Appendix A, Method 29. Such testing shall be conducted within the same time frame as the initial PM compliance testing.
- Periodic performance testing for Hg, according to 40 CFR, Part 60, Appendix A, Method 29, may be required in the future upon written notification by Ecology.

Hexavalent chromium (Cr^{+6}):

Shall not exceed the acceptable source impact level (ASIL) given in WAC 173-460-150.

Associated testing conditions:

- Testing for and compliance with the Cr^{+6} emission limit shall only be required when combusting medical waste.
- Initial performance testing for Cr^{+6} emissions at the incinerator exhaust stack and modeled to the edge of Diary Road shall be conducted according to 40 CFR, Part 60, Appendix A, Method 29 and TSCREEN modeling software. Such testing shall be conducted within the same time frame as the initial PM compliance testing.
- Periodic performance testing for Cr^{+6}, according to 40 CFR, Part 60, Appendix A, Method 29 may be required in the future upon written notification by Ecology.

Nickel (Ni):

shall not exceed the ASIL given in WAC 173-460-150.

Associated testing conditions:

- Testing for and compliance with the Ni emission limit shall only be required when combusting medical waste.
- Initial performance testing for Ni emissions at the incinerator exhaust stack and modeled to the edge of Diary Road shall be conducted according to 40 CFR, Part 60, Appendix A, Method 29, and TSCREEN modeling software. Such testing shall be conducted within the same time frame as the initial PM compliance testing.
- Periodic performance testing for Ni, according to 40 CFR, Part 60, Appendix A, Method 29 may be required in the future upon written notification by Ecology.

Establishment of operating parameters

During the initial performance tests, WSU shall establish, separately for both medical and pathological waste, the appropriate maximum and minimum operating parameters listed below, for each control system, as site-specific operating parameter

- Maximum flue gas temperature.
- Minimum secondary chamber temperature.
- Minimum horsepower or amperage to each wet scrubber.
- Minimum pressure drop across each wet scrubber.
- Minimum liquor flow rate of each wet scrubber.
- Minimum liquor pH of each wet scrubber.

Emission control monitoring

WSU shall install, calibrate, maintain and operate a CEMS for measuring opacity from the incinerator exhaust stack to the atmosphere.

- The opacity CEMS shall follow the US EPA requirements of 40 CFR 60.13, as specified in the Operations and Maintenance (O&M) manual (described below) to be submitted to, and approved of by Ecology.
- The opacity CEMS shall meet the QA requirements delineated in EPA document EPA 340/1-86-010, Recommended Quality Assurance Procedures of Opacity CEMS, or Ecology approved alternative.

- Other instrumentation for the incinerator and the scrubbers shall be fully described in the O&M manual for the facility.

Manuals

- Site-specific O&M manuals for all equipment that has the potential to affect emissions to the atmosphere shall be developed and followed. The O&M manual shall, at a minimum include the incinerator's O&M. Manufacturer's instructions should also be incorporated.
- The O&M manual shall be updated to reflect any modifications of the plant or operating procedures. Emissions that result from failure to follow the requirements of the O&M manual or manufacturer's instructions may be considered proof that the equipment was not properly operated, maintained, and tested. The O&M manual shall at a minimum include:

Normal operating parameters for the control systems

- A maintenance schedule,
- Monitoring and recordkeeping requirements,
- A description of the monitoring procedures,
- Actions for abnormal control system operation, and
- Procedures for the proper handling and disposal of incinerator ash.

Initial notifications & submittals

The requirements of the following notifications and submittals were specified by the Washington State Department of Ecology Regional Air Quality Section, Spokane, Washington (WA) 99205-1295 USA.

- *Notification of construction commencement*: WSU shall provide written notification to Ecology of the date construction of the incinerator commenced, postmarked no later than 30 calendar days after such date.
- *Notification of anticipated initial startup*: WSU shall provide written notification to Ecology of the date of initial startup of the incinerator, postmarked not more than 60 calendar days or less than 30 calendar days before such date.
- *Notification of actual date of initial startup*: WSU shall provide written notification to Ecology of the actual date of initial startup of the incinerator, postmarked within 15 calendar days after such date.
- *Notification of performance testing*: WSU shall provide written notification to Ecology of their intent to conduct any performance test at least 30 days before such test is scheduled to begin.
- *Notification of inability to conduct performance test*: If WSU is unable to conduct any performance test as scheduled, ecology shall be notified at least 24 h before the test at the above address or via phone.
- *Submittal of performance test plan*: A written test plan, including a description of the method(s) proposed, shall be submitted for approval to Ecology at least 30 calendar days prior to the start of any performance test.
- *Notification of O&M manual completion*: WSU shall provide written notification to Ecology of their having completed the O&M manual within 90 days of initial startup of the incinertor.
- *Submittal of initial site-specific operating parameters*: WSU shall provide the operating parameters identified in approval conditions no later than 60 days following initial performance testing.

Monitoring and recordkeeping

- Specific records shall be kept by WSU and made available for inspection by Ecology upon request. The records shall be organized in a readily accessible manner and cover a minimum period of the last 60 months.
- WSU shall install, calibrate (to manufacturers' specifications), maintain, and operate devices (or establish methods) for continuously monitoring, and recording as noted, the following operating parameters, except during periods of startup and shutdown:
  - Waste type and charge rate (record once per hour).
  - Maximum flue gas temperature (record once per minute).
  - Minimum secondary chamber temperature (record once per minute).
  - Minimum horsepower or amperage to each of the wet scrubbers (record once per minute).
  - Minimum pressure drop across each of the wet scrubbers (record once per minute).
  - Temperature at the outlet of each of the wet scrubbers (record once per minute).
  - Minimum liquor flow rate of each of the scrubbers (record once per minute).
  - Minimum liquor pH of each of the scrubbers (record once per minute).
  - WSU shall obtain monitoring data at all times during incinerator operation except during periods of monitoring equipment malfunction, calibration, or repair. At a minimum, valid monitoring data shall be obtained for 75% of the operating hours per day and for 90% of the operating days per calendar quarter that the facility is combusting waste.
  - WSU shall identify calendar days on which the operating parameter data were not obtained with an identification of the operating parameters not measured, reasons for not obtaining the data, and a description of corrective actions taken.
  - Nature and details of any emergency or other situation (date/time, duration, cause, etc.) and that includes situations where the incinerator was operated while the air pollution control equipment was not functioning properly.
  - WSU shall identify the calendar days and times on which the operating parameters data exceeded the applicable limits, per ecology approval conditions 8.1.1–8.1.8, with a description of the exceedances, reasons for such exceedances, and a description of corrective actions taken.
  - WSU shall maintain records of opacity measurements as determined by the CEMS.
  - WSU shall install, calibrate (to manufacturers' specifications), maintain, and operate a device or method for measuring the use of the bypass stack including the date, time, and duration.
  - Waste throughputs by type (medical, low-level radioactive, and pathological) and quantity (total for each calendar year).
  - Hours of incinerator operation and the quantity of Natural gas usage (total for each calendar year).
  - A file of initial and any periodic performance testing results.
  - O&M manual and maintenance records.
  - WSU shall keep a record of the incinerator operators that are certified per WAC, along with their certification and renewal dates.

Reporting

The following reports shall be sent, within 45 days following the end of the calendar year unless otherwise noted below to the Washington State Department of Ecology:

- The site-specific operating parameters identified or established by subsequent performance testing.
- The highest operating parameters and the lowest operating parameters for each parameter identified at approval conditions.
- The highest operating parameters and the lowest operating parameters for each parameter identified for the calendar year preceding the year being reported. This will provide a summary of the performance of the facility over a 2-year period.
- A statement that no omissions, situations or exceedances occurred during the prior calendar year, if no omissions, situations or exceedances were recorded.
- The operating parameter data not obtained per the recordkeeping.
- The nature and details of any emergency or other situation per the recordkeeping.
- The operating parameters data that exceeded the applicable limits per the recordkeeping.
- Any information recorded under for the calendar year preceding the year being reported. This will provide a summary of the performance of the facility over a 2-year period.
- Instances of the use of the bypass stack per the recordkeeping.
- Waste throughputs by type (medical, low-level radioactive and pathological) and quantity per the recordkeeping.
- Hours of incinerator operation per the recordkeeping.
- Natural gas usage per the recordkeeping
- The results of all initial performance and any subsequent periodic testing shall be sent to the above address no later than 60 days following such testing.
- Semi-annual reports containing any information recorded shall be submitted to the above address no later than 60 days following the end of the reporting period.
- The first semi-annual reporting period ends 6 months following the date the initial site-specific operating parameters are submitted.
- Subsequent semi-annual reports shall be submitted no later than 6 calendar months following the previous report.

Specific conditions

- WSU shall ensure that the control equipment is not operated above any of the applicable maximum operating parameters or below any of the applicable minimum operating parameters listed under the conditions of approval. The operating parameters shall be measured as 3-hour rolling averages (calculated each hour as the average of the previous three operating hours) at all times except during periods of startup, shutdown and malfunction. Operating parameter limits do not apply during performance tests. Operation above the established maximum or below the established minimum operating parameter(s) shall constitute a violation of established operating parameter(s).
- Operation of the facility above the maximum charge rate and below the minimum pressure drop across the wet scrubber or below the minimum horsepower or amperage to the system (each measured on a 3-hour rolling average) simultaneously shall constitute a violation of the PM emission limit.
- Operation of the facility above the maximum charge rate and below the minimum secondary chamber temperature (each measured on a 3-hour rolling average) simultaneously shall constitute a violation of the CO emission limit.

- Operation of the facility above the maximum charge rate, below the minimum secondary chamber temperature, and below the minimum scrubber liquor flow rate (each measured on a 3-hour rolling average) simultaneously shall constitute a violation of the Dioxins/Furans emission limit.
- Operation of the facility above the maximum charge rate and below the minimum scrubber liquor pH (each measured on a 3-hour rolling average) simultaneously shall constitute a violation of the HCl emission limit.
- Operation of the facility above the maximum flue gas temperature and above the maximum charge rate (each measured on a 3-hour rolling average) simultaneously shall constitute a violation of the Hg emission limit.
- Use of the bypass stack (except during startup, shutdown, or malfunction) shall constitute a violation of the PM, Dioxins/Furans, HCl, Pb, Cd and Hg emission limits.
- WSU may conduct a repeat performance test within 30 days of violation of applicable operating parameter(s) to demonstrate that the facility is not in violation of the applicable emission limit(s). Such a repeat performance test shall be conducted under the identical operating parameters that resulted in an emission violation under one or more of the conditions listed at the time of approval.
- WSU may conduct a repeat performance test at any time to establish new values for the operating parameters.
- Additional air pollution controls or handling procedures may be required by Ecology in order to control odors if a nuisance is identified in the future.
- The incinerator shall not be operated without a certified operator in responsible charge on-site during all hours of operation.
- Waste shall not be charged to the incinerator during periods of start-up, shutdown, or malfunction.
- Medical waste shall be incinerated under sufficient burning conditions to reduce all combustible material to a form such that no portion of the combustible material is visible in its initial state and not combusted.
- Incinerator ash shall only be transferred from the incinerator to sealable metal containers within the enclosed incinerator building followed by disposal at an approved site.
- The use of treatment chemicals for the cooling tower shall be chemicals that are not toxic air pollutants per Chapter 173–460 WAC.

General conditions

Visible emissions: No visible emissions shall be allowed beyond the property line, as determined by opacity readings.

Commencing/discontinuing construction and/or operations: This approval shall become void if construction of the project is not commenced within 18 months after receipt of this Order approving the NOC application, or if construction or operation of the incinerator is discontinued for a period of 18 months.

Compliance assurance access: Access to the source by EPA or Ecology shall be allowed for the purposes of compliance assurance inspections. Failure to allow access is grounds for revocation of the Order approving the NOC.

Availability of order: Legible copies of the Order approving the NOC application shall be in the working vicinity of the incinerator, be available to employees in the direct operation of the systems and be available for review upon request by Ecology.

Equipment operation: Operation of the incinerator shall be conducted in compliance with all data and specifications submitted as part of the NOC application and in accordance with the O&M manual, unless otherwise approved in writing by Ecology.

Outdoor burning: As provided in WAC 173-425-040, the following materials shall not be burned in any outdoor fire at the facility: garbage, dead animals, asphalt, petroleum products, paints, rubber products, plastics, paper (other than what is necessary to start a fire), cardboard, treated wood, construction debris, metal or any substance (other than natural vegetation) which when burned releases toxic emissions, dense smoke, or odors. Contact the local fire department or the Ecology Burning Specialist, Department of Ecology before conducting outdoor burning.

Activities inconsistent with this order: Any activity undertaken by WSU or others, in a manner that is inconsistent with the NOC application and this determination, shall be subject to Ecology enforcement under applicable regulations.

Obligations under other laws or regulations: Nothing in this Order shall be construed to relieve WSU of its obligations under any local, state or federal laws or regulations.

Fees: As per this order and related regulatory requirements, the Washington State Department of Ecology will have a fee associated for review and issuance of permits/approvals.

9.4 CROSS-MEDIA TRANSFER POTENTIAL OF INCINERATION TECHNOLOGIES

9.4.1 General

The general cross-media transfer potentials during pre- and post-treatment activities are addressed in Chapter 12.

9.4.2 Additional concerns for incineration technologies

Other concerns include various types of emissions during the incineration treatment. Types of potential releases are below:

- Fugitive emissions from fuel sources can add to the overall emissions of organics from the site which have higher potential as greenhouse gas resulting in global warming.
- Incomplete combustion of organic compounds can generate "products of incomplete combustion" (PICs) which are responsible for the climate change (see Chapter 2) and can cause serious air pollution (USEPA, 1995).
- These emissions from the incinerators are increasingly adding an enormous amount of PICs, such as CO_2, NO_x, and Methane which are responsible for the greenhouse effect and cause global warming (EC, 2020).
- High combustion gas flow can cause problems in the gas cooling device (Vickery, 1995).
- Scrubber gases are often very acidic (when the combusted soils contain high levels of chlorine or sulfur) and may cause corrosion in the system components, which in turn may cause containers to leak/fail, resulting in potential cross-media transfer of contaminants.
- Potential risk of metal emissions with dust or due to volatilization (e.g., mercury).
- Low pH of scrubber water could cause equipment damage and leaks.

- Unless effectively managed and controlled, scrubber water/residuals from air pollution control devices also pose some threat for the release of contaminants to the air, water, or soils.
- Residuals (water) generated from waste conditioning before incineration.
- There is potential for VOC emissions due to leaks in pipes, joints, valves, and uncovered conveyor systems.
- Stack emissions from the destruction of organic vapors have the potential to release contaminants in the air/natural environment, above the regulatory limit.
- Inadequate control and management of baghouse dust (including ash, metals, and/or unoxidized compounds) may cause transfer of contaminants to the environment.
- When the internal temperature and/or pressure of the combustion chamber become too high, emergency vents are opened to allow some of the superheated gases to escape. This may allow untreated air containing volatiles, heavy metals, or other substances to escape, leading to the potential risk of polluted air emissions.

9.5 BEST MANAGEMENT PRACTICES/OPTIONS (BMPs) FOR INCINERATION TECHNOLOGIES TO AVOID POTENTIAL CROSS-MEDIA TRANSFER OF CONTAMINANTS

General BMPs to prevent potential cross-media transfer of contaminants during pre- and post-treatment activities are discussed in Chapter 12. The technology-specific treatment activities and the possible BMP options to control specific cross-media transfer of contaminants during these activities are discussed in this subsection.

During the incineration process, the following activities are most commonly undertaken:

- The contaminated media is moved through a conveyor system to the incineration chamber. Heat is applied in the chamber and the waste subjected to flame oxidation/burning at a high temperature in a turbulent environment for a period of time necessary to convert the waste into carbon dioxide and water. Rotary or other types of mixing actions are generally undertaken. Waste heat is supplied to the boiler which supports the air and water pollution control units.

The following BMPs are generally recommended, where appropriate, for preventing the cross-media transfer of contaminants when the above activities are conducted:

- During the main treatment activities organic or inorganic vapor emissions should be monitored and appropriate emission control measures – described in Chapter 13 of this book – be employed to prevent emissions above the allowable level specified by the regulatory agency (EPA or the authorized state).
- The fuel storage and fuel handling areas should be included under monitoring and emission control oversight, if deemed necessary.
- Pipes, valves and fittings – where fuel or pressurized liquids are involved – should be checked regularly for leaks. The general condition of the joints should also be checked.
- Three major waste streams are generated by this technology: solids from the incinerator and air emissions control system; water from the air emissions control system; and emissions from the incinerator. The following BMPs should be used to control the potential of cross-media contamination from the management of these residuals:

- o Ash and treated soils or solids from the incinerator combustion chamber, as well as solids from the air emissions control system, such as fly ash or granulated activated carbon (GAC), may be contaminated with heavy metals. These residues should be tested with leachate toxicity tests. If they fail these tests, they should be treated by a process such as solidification/stabilization (see Chapter 10) and disposed of onsite or in an approved landfill (Freeman and Harris, 1995).
- o Liquid waste from the air pollution emissions system may contain caustics, high levels of chlorides, volatile metals, trace organics, metals, and inorganic particulates. Treatment may require neutralization, chemical precipitation, evaporation, filtration, or carbon adsorption before discharge (Freeman and Harris, 1995).

9.6 WASTE CHARACTERISTICS IN INCINERATION TECHNOLOGIES THAT MAY INCREASE THE CHANCES OF CROSS-MEDIA CONTAMINATION

The effectiveness of incineration treatment technologies could be compromised and undue cross-media contamination may be caused under certain conditions as identified in this subsection. However, some of these limitations could possibly be overcome with certain technology-specific modifications and variations. Please refer to technology-specific references provided at the end of this chapter for additional information about modifications or variations that can be used to enhance the effectiveness of an incineration technology (ASME, 1988; Dutta, 2002; USEPA, 1988, 1990, 1993). Some of the conditions impacting the effectiveness of incineration treatment technologies are listed below:

- Since incineration does not destroy most inorganic (metals) wastes, this treatment technology may not be effective for waste media containing metals.
- It may be fuel-intensive for wastes with high moisture content.
- Wastes with low organic content may be costly to incinerate.
- Some explosive wastes may require a specially designed incinerator.
- This technology will not be applicable to wastes requiring in situ treatment.
- Wastes with high debris/large particle content may be a problem for some incinerators due to the size limitation of the feeder door.

REFERENCES

ASME. (1988). *Hazardous Waste Incineration, A Resource Document.* American Society of Mechanical Engineers, New York, January.

Dutta, S. (2002). *Environmental Treatment Technologies for Hazardous and Medical Wastes, Remedial Scope and Efficacy.* Tata McGraw Hill Publishing Company, New Delhi.

European Commission (EC). (2020). *Causes of Climate Change.* [Online] Available from: https://ec.europa.eu/clima/change/causes_en [Accessed 17th October, 2020].

Freeman, H. M., and E. F. Harris. (1995). *Hazardous Waste Remediation: Innovative Treatment Technologies.* Technomic Publishing Co., Inc., Lancaster, PA.

Honican, J., and S. Satterlee. (1998). Letter to the Department of Ecology, State of Washington, 4601 N. Monroe, Suite 202, Spokane, WA 99205-1295, Concerning Installation of a new Incinerator at the Washington State University (WSU), April 17.

USEPA. (1988). *Hazardous Waste Incineration: Questions and Answers*. Office of Solid Waste, Washington, DC. EPA/SW-88-018.

USEPA. (1989). *Seminar Publication Corrective Action Technologies and Applications*, Office of Solid Waste, EPA/625/4-89/020, pp. 41–47.

USEPA. (1990). *Engineering Bulletin-Mobile/Transportable Incineration Treatment*. Office of Research and Development, Cincinnati, OH. EPA/540/2-90/014, September.

USEPA. (1993). *Approaches for the Remediation of Federal Facility Sites Contaminated with Explosive or Radioactive Wastes*. Office of Research and Development, Washington, DC. EPA/625/R-93/013, September.

USEPA. (1994). *Innovative Site Remediation Technology-Thermal Destruction*. Vol. 7. Office of Solid Waste and Emergency Response. EPA/542/B-94/003, October.

USEPA. (1995). *BMP Development Workshop Summary-Incineration Technologies*. Summary of Workshop on Incineration Technologies, March.

USEPA. (1997). *Best Management Practices (BMPs) for Soil Treatment Technologies*. Office of Solid Waste, Washington, DC. EPA-530-R-97-007, May.

USEPA. (2012). A *Citizen's Guide to Incineration*. Washington, DC. EPA 542-F-12-010, September, .

Vickery, R. (1995). *Personal Communication*. Memo from Dupont Facilities Services, Wilmington, NC, Delaware, to Subijoy Dutta, October 13, 1995.

Washington State Department of Ecology. (2020). Regulations & Permits > Permits & Certifications > Air Quality Permits > Notice of Construction permit. [Online] Available from: https://ecology.wa.gov/ [Accessed August 25th, 2020].

Chapter 10

Other physical/chemical treatments

This chapter focuses on remedial actions using "other physical/chemical treatment technologies".

This category includes all technologies that are not within the purview of the broad categories of treatment technologies covered in the previous chapters. *Most of the in situ treatment* technologies are discussed in this chapter. The best management practices/options (BMPs) presented here can be applied when appropriate and as it best fits the site-specific applications of the remedial treatment.

10.1 DEFINITION AND SCOPE OF OTHER PHYSICAL AND CHEMICAL TREATMENT

The classification of "Other Physical and Chemical" (OPC) treatment envelops a variety of novel treatment technologies that are currently being applied to treat contaminated soil and other media. These technologies can be applied independently or in conjunction with other methods to enhance removal and/or stabilization of the contaminants (USEPA, 1997).

The following are a few examples of OPC treatment technologies:

- In situ radio frequency heating
- In situ vitrification
- In situ IR heating
- In situ redox control
- In situ soil flushing
- Solidification/stabilization
- Fracturing
- Solvent extraction
- Phyto uptake (wetlands)
- Gasification/pyrolysis
- Debris washing
- Grouting
- Dechlorination

The scope of other physical/chemical treatment is not limited to the above-listed technologies. Although a number of treatment technologies are listed within this category, the ones discussed here in detail are in situ radio frequency (RF) heating, in situ vitrification (ISV), in situ soil flushing, solidification/stabilization (S/S), and off-site disposal because of their broad applicability.

DOI: 10.1201/9781003004066-10

10.2 IN SITU RADIO FREQUENCY HEATING

10.2.1 Definition and scope of in situ RF heating

RF heating technologies use electromagnetic energy generated by radio waves to heat soil in situ, thereby potentially enhancing the performance of standard soil vapor extraction (SVE) technologies. RF heating is designed to accelerate the removal of volatile organics and to make it possible to remove semi-volatile organics that would not normally be removed by standard SVE technologies. Contaminants are removed from in situ soils and transferred to collection or treatment facilities. The RF energy causes dielectric heating of the soil. Some conductive heating also occurs in the soil (USEPA, 1995a).

RF heating is applicable to wastes located above the water table, only when the saturated soils, if any, can be effectively dewatered.

A schematic in situ RF heating system is shown in Figure 10.1, and a cross-sectional diagram is shown in Figure 10.2 (USEPA, 1995a).

10.2.1.1 Key features of In situ RF heating

- Enhances the ability of SVE systems to remove organic contaminants.
- Involves in situ installation of vapor extraction wells and electrodes or antennae.
- Employs a dielectric heating frequency between 2 and 2,450 MHz. Operating on a frequency band allocated for industrial, scientific, and medical (ISM) equipment.

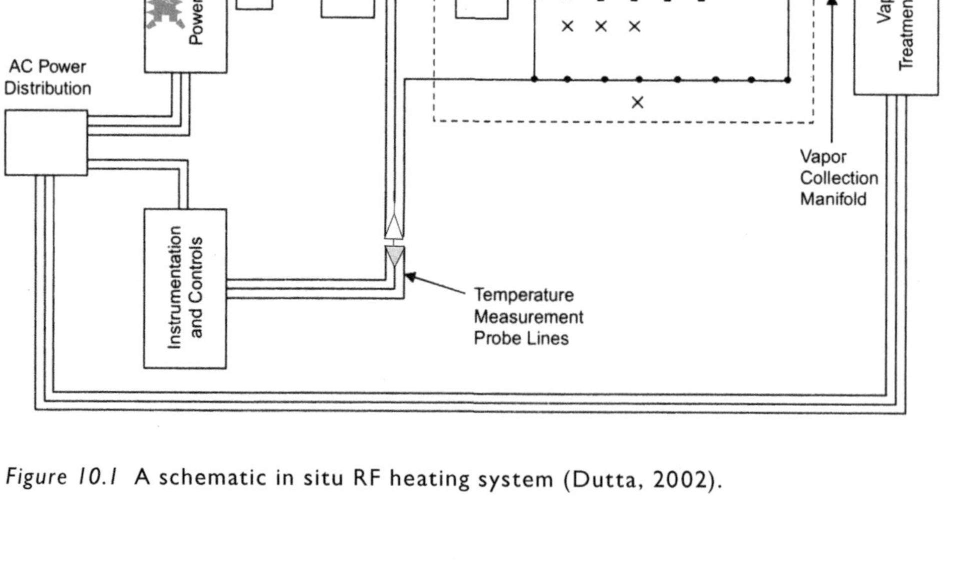

Figure 10.1 A schematic in situ RF heating system (Dutta, 2002).

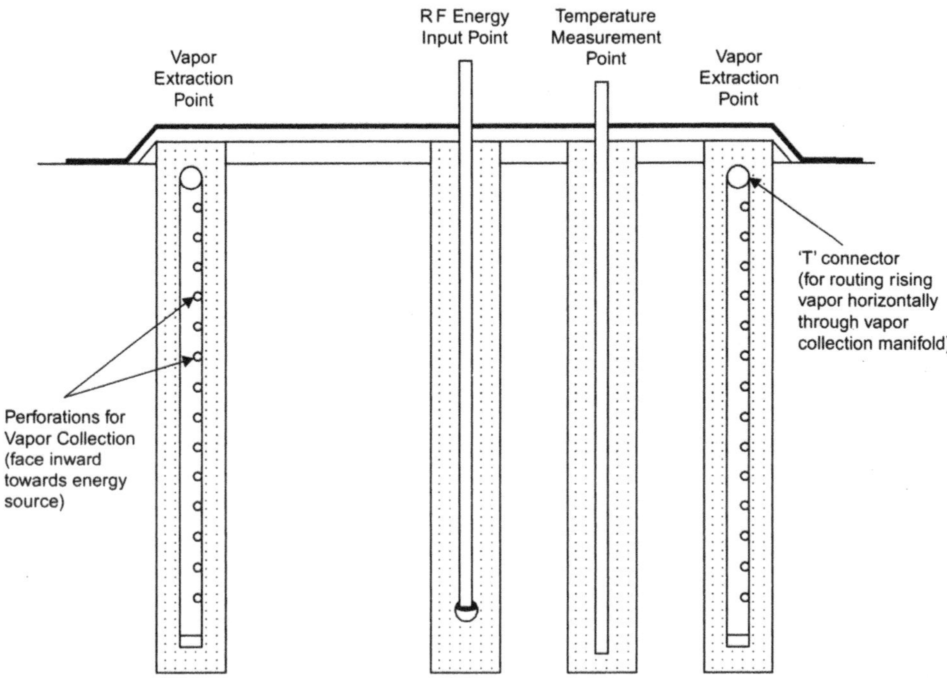

Figure 10.2 Cross-section of an in situ RF heating system (Dutta, 2002).

- Heat soils to a temperature range of 100°C–300°C (212°F–572°F), on average for the treated area.
- Field applications involve the installation of:
  - Boreholes into the contaminated area
  - A vapor extraction and treatment system
  - RF shield (sometimes) to control RF emissions
  - Electrodes or antennae.

10.2.2 In situ RF heating technology description

RF heating is performed by applying electromagnetic energy in the RF band. The energy is delivered to electrodes placed into the soil cover. The mechanism of heat generation is similar to that of a microwave oven and does not rely on the thermal properties of the soil matrix. The power source for the process is a modified radio transmitter. The exact frequency of operations is selected after the evaluation of the dielectric properties of the soil matrix and the size of the area requiring treatment. As the soil is heated due to the dissipation of the RF energy, contaminants and moisture in the soil are vaporized and pulled toward ground electrodes, which also serve as vapor extraction wells. The vaporized water may act as a steam sweep to further enhance the removal of organic contaminants. A standard SVE system provides a vacuum to the ground electrodes and transfers the vapors to collection or treatment facilities. Contaminants are treated using standard vapor treatment techniques. After soil treatment is complete, the soil is allowed to cool. The SVE system may be operated during part or all of this cooling period. The exact number of exciter and ground electrodes, electrode configurations, vapor collection or treatment techniques, and other design details are generally site-specific (Dutta, 2002).

10.2.3 Cross-media transfer potential of in situ RF heating

10.2.3.1 General

General cross-media transfer potentials during pre- and post-treatment activities are addressed in Chapter 12. In addition, the cross-media concerns described for vapor extraction (Chapter 7) are generally applicable to this technology too.

10.2.3.2 Additional concerns for in situ RF heating

- Release of contaminants (VOCs and particulate matters) during site preparation and borehole drilling.
- Migration of hot vapors to cooler zones and the resulting re-condensation of contaminants.
- Surface water intrusion from beyond the boundaries of the off-gas hood into the contaminated area that is being treated.
- Downward migration of contaminant vapor condensates to aquifer.
- Seasonal variation of the water table, which causes contaminants to move as the saturated zone fluctuates.
- The unanticipated occurrence of inorganics, such as mercury, can increase requirements for system operation, emissions control, and disposal of off-gases or other residues.
- Unexploded ordnance (UXOs) may pose a potential problem.
- With the application of an SVE system, air can move through preferred channels, such as natural gravel lenses or fill materials around buildings or utility lines. This can result in ineffective cleanup, leaving areas untreated or treated to levels higher than allowable for closure.

10.2.4 Best management options for controlling the cross-media transfer potential of in situ RF heating

General BMPs to prevent potential cross-media transfer of contaminants during pre- and post-treatment activities are discussed in Chapter 12. However, BMPs to control specific cross-media transfer of contaminants due to the specific activities during the in situ RF heating treatment process are discussed in this subsection.

The activities most commonly undertaken during the RF heating treatment process are summarized below:

- *Power source operation*: In these systems, the temperature rise occurs as a result of ohmic or dielectric heating mechanisms. Ohmic heating arises due to the ionic current or conduction current that flows in materials in response to the applied electric

field. Dielectric heating arises from the physical distortion of the atomic or molecular structure of polar materials in response to the applied electric field. Since the electric field changes direction rapidly in the RF heating, the alternating physical distortion dissipates mechanical energy that is thermal energy in the material. The RF heating process raises the temperature of the soil to a range between 100°C and 300°C. It employs an array of metal electrodes, which are placed in boreholes drilled through the contaminated soil. The ground electrodes are generally supplied with 480-V, 3-phase power in major applications.
- *Vapor collection and treatment*: The hot gases and vapors are collected by means of a gas collection system and transported to the on-site vapor treatment system by means of a vacuum blower. If carbon adsorption is used to treat vapor, compressed air may be used for system control. Steam or hot air is supplied when the carbon bed is regenerated on-site. Natural gas or propane is used if flare is used to control vapors.

To prevent cross-media transfer of contaminants while carrying out the above activities, the following BMPs are generally recommended to be used, when necessary:

- Soil permeability should be tested to determine whether the vapor extraction system will be capable of efficiently collecting the volatilized contaminants and to optimize the location of vapor extraction wells.
- The site should be characterized in terms of the contaminants present, particularly any volatile metals. This assists in determining the contaminants which could be volatilized by RF heating and collected by the vapor extraction system.
- Buried metallic objects such as drums and tanks should be removed to the extent possible. In addition, the presence of buried explosive materials should be checked to eliminate explosion potential.
- RF excitor electrodes should not be installed close to a building or any other structure.
- The configuration, number, depth, and orientation of extraction wells should be determined so that the vapor extraction system can efficiently collect volatilized contaminants with no zones of stagnant air.
- The well diameter, length, and location of screened zones of extraction wells should be optimized so that the SVE system can efficiently collect volatilized contaminants.
- Extraction wells should not be located near surface water impoundments, underground storm sewers, or drains.
- Well casings, screens, and other structural materials should be protected from the potential effects of RF heating, such as the development of leaks or cracks.
- Leaks and preferential pathways should be checked following the completion of construction to ensure that the SVE system can efficiently collect all volatilized contaminants.
- The vapor barrier system should be designed to ensure that none of the contaminants volatilized by the RF heating is released into the ambient air through the surface.
- During the main treatment activities, as specified above, RF monitoring should be conducted to ensure that the RF field outside of the treatment zone and within operation control areas does not exceed the National Institute of Occupational Safety and Health (NIOSH) or Federal Communications Commission (FCC) requirements.
- Organic or inorganic vapor emissions should be monitored, and appropriate emission control measures, as described in Chapter 13, should be employed to prevent emissions above the allowable level specified by the regulatory authority.

- If the vapor treatment system yields any moisture or liquid from the treatment process, the contaminated aqueous stream should be collected in a tank or a lined/contained system. This will prevent the contaminants from mixing with the normal surface water runoff from the area and the surrounding natural watercourse. The contaminated aqueous stream should be treated or disposed of in accordance with the applicable regulations.
- Continued use of emission control systems may be necessary after the treatment is completed since RF heating elevates the soil temperature and volatiles may continue to be emitted for some time following the treatment.

10.2.5 Waste characteristics that may increase the chances of cross-media contamination in any in situ RF heating

The effectiveness of in situ RF heating technologies could be compromised, and cross-media contamination may be caused under certain conditions identified in this subsection. However, some of these limitations could possibly be overcome with technology-specific modifications and variations. Please refer to technology-specific references provided at the end of this chapter for additional information about modifications or variations that can be used to enhance the effectiveness of in situ RF heating technologies (Dev, 1995; Dutta, 2002; USEPA, 1995c, 1995b, 2012a).

- Highly saturated (greater than 25% moisture) soils will consume enormous amounts of energy as the application of the RF heat causes the water to evaporate, reducing the efficiency of the RF system.
- RF heating will not treat nonvolatile organics, inorganics, metals, and heavy oil.
- Certain RF heating technologies may not operate most effectively at depths below 50 feet (USEPA, 1995a).
- Contaminants in clayey soils are usually strongly sorbed and difficult to remove (USEPA, 1995b). This reduces the efficiency of the RF system.
- In situ RF heating treatment involves drilling equipment and other heavy machinery to install wells or electrodes. These activities cause increased traffic, noise, dust emissions, and possibility of vapor leaks. Because of these issues, the in situ RF heating technologies are not generally recommended for use nearby residential areas (USEPA, 2012a).

10.3 IN SITU VITRIFICATION

10.3.1 Definition and scope of in situ vitrification (ISV)

In situ vitrification (ISV) uses electrical power to heat and melt earthen materials (e.g., soils, sludges, mine tailings, and sediments), waste materials buried in earthen materials, and other earthen-like materials contaminated with organic, inorganic, and metal-bearing hazardous and/or radioactive wastes. The molten material cools to form a hard, monolithic, chemically inert, amorphous or crystalline product that incorporates and immobilizes the thermally stable inorganic compounds and heavy metals in the hazardous waste. Organic contaminants are either volatilized or captured at the hood or pyrolyzed for more volatilization of remaining constituents and subsequent capture at the hood. The slag product is a glass-like material with very low leaching characteristics.

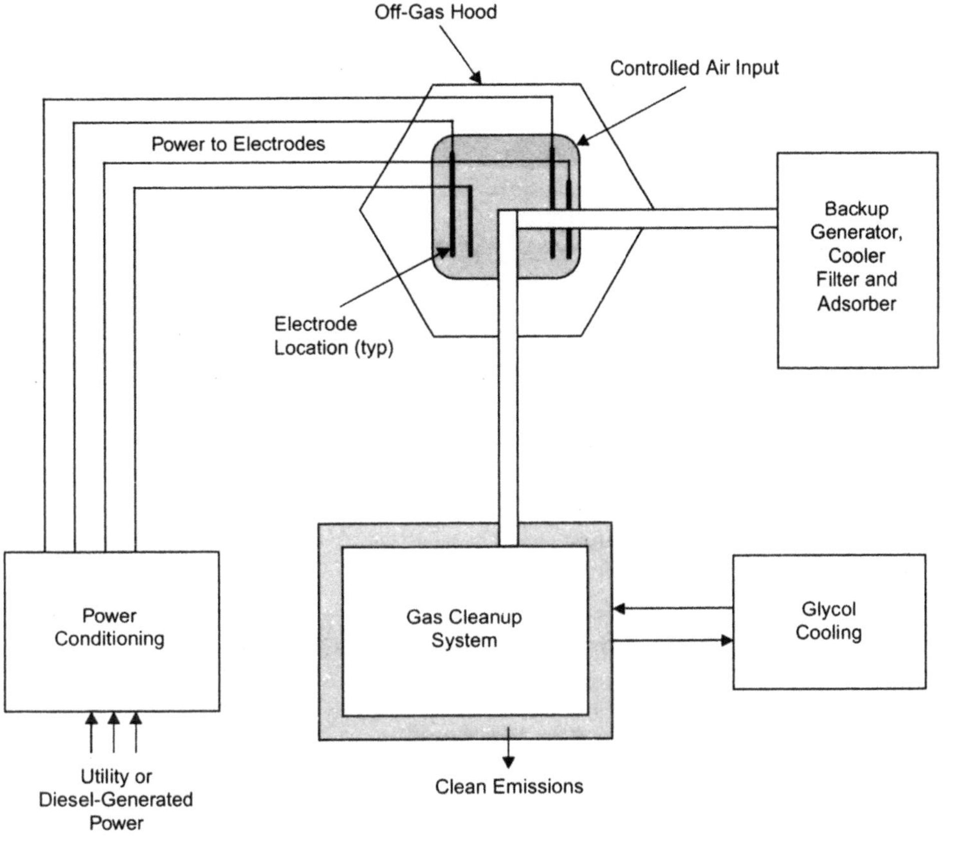

Figure 10.3 ISV equipment system (Dutta, 2002).

A flowchart of the ISV process is shown in Figure 10.3. A cut-away view of a treatment cell specifying the general limits for the volume treated is shown in Figure 10.4 (Dutta, 2002).

10.3.1.1 Key features of ISV (Dutta, 2002)

- Transforms non-volatile inorganic waste into a non-leachable vitrified mass.
- Gases/vapors are passed through the off-gas treatment system and released thereafter to the environment.
- Uses electrical energy to heat and melt soil.
- Heats soils to a temperature range of 1,600°C–2,000°C (2,900°F–3,600°F).
- Field applications involve the installation of:
  - o 12-inch outside diameter (OD) graphite electrodes
  - o Off-gas quenching system that includes a water-based quenching tower and high-efficiency Venturi scrubber; a secondary cooling system (glycol-based) is then employed to keep the water temperature within limits
  - o Optional aboveground thermal oxidizer for treating organic vapors
  - o Electrical supply line and transformer.

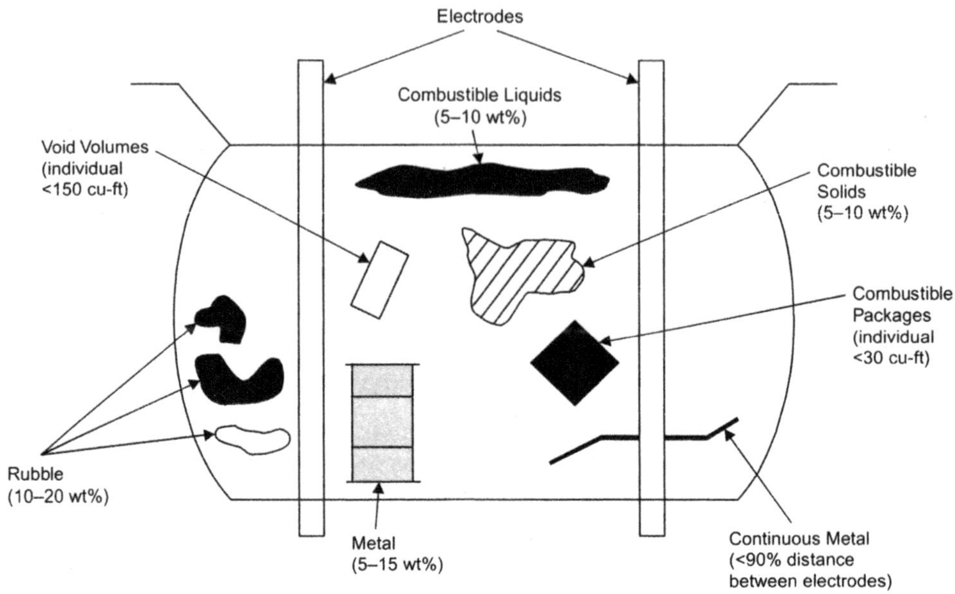

Figure 10.4 A cut-away view of a typical treatment cell (Dutta, 2002).

10.3.2 In situ vitrification technology description

Several methods and configurations exist for the application of ISV. At a site that only has a relatively shallow layer of contamination, the contaminated layer may be excavated and transported to a pit where the vitrification will take place. At other sites where the contamination is much deeper, thermal barriers could be placed along the ISV site and prevent the movement of heat and glass into adjacent areas. This will force the heat energy downward, and melt depths will be increased. A more conventional approach to using ISV is to encapsulate the wastes and control the lateral migration of contaminants within a checkerboard pattern of melts.

ISV uses a square array of electrodes up to 18 feet (5.5 m) apart, which is inserted to a depth of 1–5 feet (0.3–1.5 m) and potentially can treat down to a depth of 20 feet (6 m) to remediate a contaminated area. As the vitrified zone grows, it vitrifies metals and either vaporizes or pyrolyzes organic contaminants. A hood placed over the processing area is used to collect combustion gases, which are treated in an off-gas treatment system. The gases collected by the hood are treated by quenching, scrubbing, mist elimination, heating, particulate filtration, and activated carbon adsorption (USEPA, 1994a).

10.3.3 Cross-media transfer potential of ISV

10.3.3.1 General

General cross-media transfer potentials during pre- and post-treatment activities are discussed in Chapter 12.

10.3.3.2 Additional concerns for ISV

Other concerns include various types of emissions during the ISV treatment. Types of potential releases are listed below:

- Release of contaminants (volatiles and particulates) during site preparation.
- Possible hazards due to high voltage electrical fluxes.
- Downward migration of melted contaminants or contaminant vapors and condensates to aquifer.
- Contaminant vapors can be released away from the treatment zone if there is an open pathway (e.g., pipe and French drain) that intercepts the treatment zone. Sites should be inspected for open subsurface conduits before applying ISV.

10.3.4 Best management practices/options for controlling cross-media transfers during ISV

During the ISV treatment activities, steps may need to be taken to control fugitive emissions and to prevent the release of contaminated media to the natural environment. The following BMPs are generally recommended for preventing cross-media transfer of contaminants when the above activities are being carried out:

- During large-scale operations, pumping wells and/or intercept trenches should be installed around the treatment zone. This will prevent groundwater from flowing through a contaminated treatment zone.
- If open subsurface conduits are found, they should be disrupted (e.g., collapsed, broken, and filled) before the ISV treatment.
- The ISV treatment activities should be conducted under a controlled environment when there are possibilities of the off-gases, volatiles, dust, etc. emissions at levels above the regulatory limit. The VOC emissions associated with these activities should be controlled by capturing these emissions and then treating the captured vapor/air.
- To minimize the possibilities of contaminated soil from being airborne, a thin layer of clean cover soil should be placed over the contaminated area before the placement of the off-gas collection hood. This practice helps to keep the contamination within the collection hood and tracking of contaminated material across the site.
- If the vapor treatment system yields any moisture or liquids from the treatment process, the contaminated aqueous streams should be collected in a tank or a lined/contained system. This will prevent the contaminants from mixing with the normal surface water runoff from the area and the surrounding natural watercourse. The contaminated aqueous stream should be treated or disposed of in accordance with the applicable regulations.
- Typically, 20%–40% volume reductions occur as a result of the melting process, which can lead to subsidence within the treated zone. Clean soil is generally placed to fill in the subsided region as a standard practice, and this should be consistently followed in all ISV treatments.
- Arrangements should be made for leaving the vitrified mass in place or exhuming it and disposing of it properly.
- If the vitrified soil is to be left in place, clean soil should be backfilled over it to encourage the regrowth of vegetation.

- Electrodes may be removed, decontaminated, reused, disposed of, or left in the vitrified mass.
- Drums or tanks filled with liquid, if removed from the treatment area, should be disposed of in compliance with the applicable state and/or federal regulations.

10.3.5 Waste characteristics that may increase the likelihood of cross-media contamination for ISV technologies

The effectiveness of ISV technologies could be compromised and undue cross-media contamination may be caused under certain conditions, as given below:

- Organic contents more than 10% by weight within the soil matrix might require additional or modified off-gas treatment components. Alternatively, soils of higher organic contents have been successfully treated by slowing the melting rate to levels that allow acceptable heat removal rates in the off-gas treatment system (Campbell, 1995).
- Pockets of flammable liquid or vapor sealed in containers beneath the soil surface can create a potential explosion hazard.
- In cases where heavy metal immobilization is desired, soils containing less than 50% by weight of glass formers (e.g., aluminum and silica) and <2% by weight of alkali compounds (e.g., sodium and potassium) may require modification with additives to obtain desired melt and vitrified product characteristics (Campbell, 1995).
- Depths greater than 20 feet (6 m) require higher power-level equipment because of the larger masses and volume that are being treated (Campbell, 1995).
- Soils that contain inorganic debris greater than 55% by volume are extremely difficult to treat with this technology (Campbell, 1995).
- Soils with high groundwater recharge rates require special methods to limit recharge to acceptable rates.

However, some of these limitations could possibly be overcome with certain technology-specific modifications and variations. Please refer to technology-specific references provided at the end of this chapter for additional information about modifications or variations that can be used to enhance the effectiveness of ISV technologies (Dutta, 2002; USEPA, 1990, 2012a).

10.4 IN SITU SOIL FLUSHING

10.4.1 Definition and scope of in situ soil flushing

In situ soil flushing is the extraction of contaminants from the soil with water or other suitable aqueous solutions. Soil flushing is accomplished by passing the extraction fluid through in-place soils using an injection or infiltration process. Extraction fluids must be recovered and, when possible, recycled. Typically, soil flushing is used in conjunction with other treatments that destroy contaminants or remove them from the extraction fluid and groundwater (USEPA, 1992).

In situ soil flushing includes conventional and unconventional techniques. The conventional techniques include well-and-capture methods in the vadose zone and pump-and-treat systems in the saturated zone. Unconventional techniques consist of primary, secondary, and tertiary recovery techniques (USEPA, 1993a).

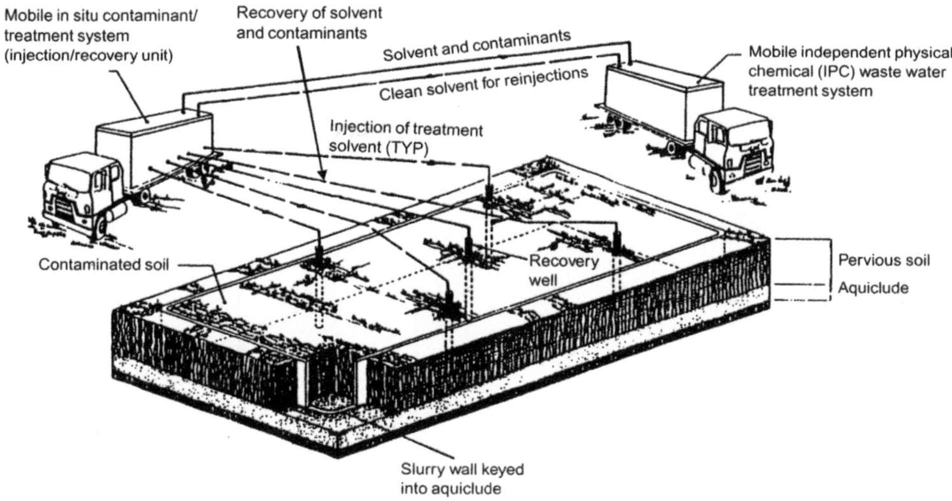

Figure 10.5 A schematic in situ flushing system used in a field test (Dutta, 2002).

Figure 10.5 shows a schematic of an in situ soil flushing system in which the treatment solvent is injected into the soil. Figure 10.6 is a schematic of an in situ soil flushing sprinkler system. Figure 10.7 is a cross-section of an in situ soil flushing system that uses spray application (USEPA, 1991).

10.4.1.1 Key features of in situ soil flushing (Dutta, 2002)

- Fluid injection
- Contaminant mobilization and removal.
- Secondary and tertiary recovery in some cases.
- Field applications involve installation of:
  - o Subsurface injection wells or aboveground sprinkler/infiltration bed systems
  - o Boreholes for recovery wells or other subsurface recovery devices in the contaminated area
  - o Delivery and recovery drain lines
  - o Reagent delivery system
  - o Produced fluid treatment system
  - o Physical (e.g., sheet pile wall) or hydraulic (e.g., groundwater depression or mounding) barriers to contain contaminants.

10.4.2 In situ soil flushing technology description

Schematics of different types of in situ soil flushing applications are provided in Figures 10.5–10.7.

In situ soil flushing includes conventional and unconventional techniques. The conventional techniques include well-and-capture methods in the vadose zone and pump-and-treat

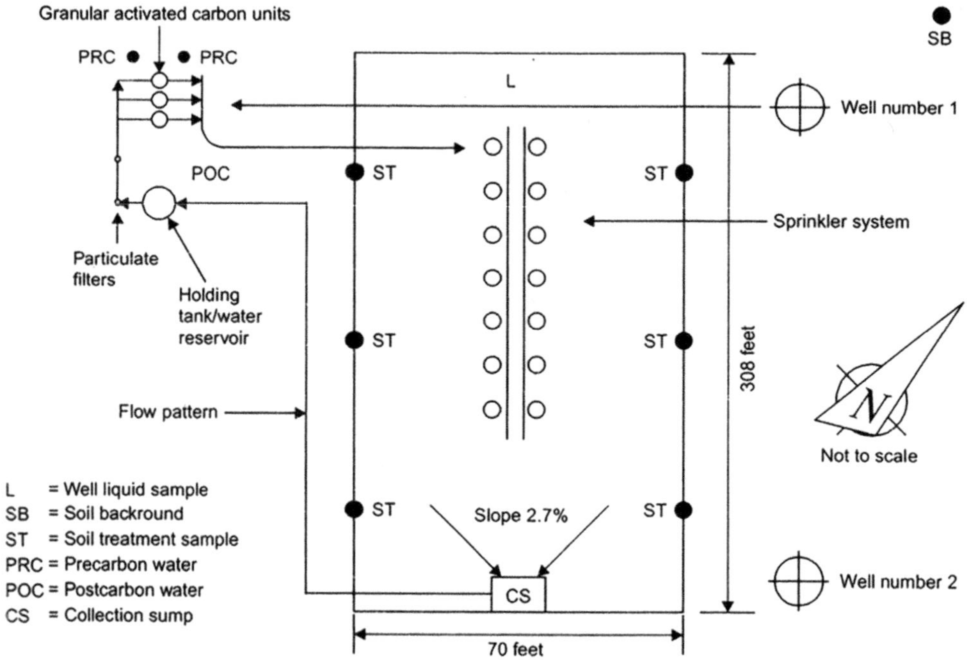

Figure 10.6 Schematic soil flushing sprinkler system (Dutta, 2002).

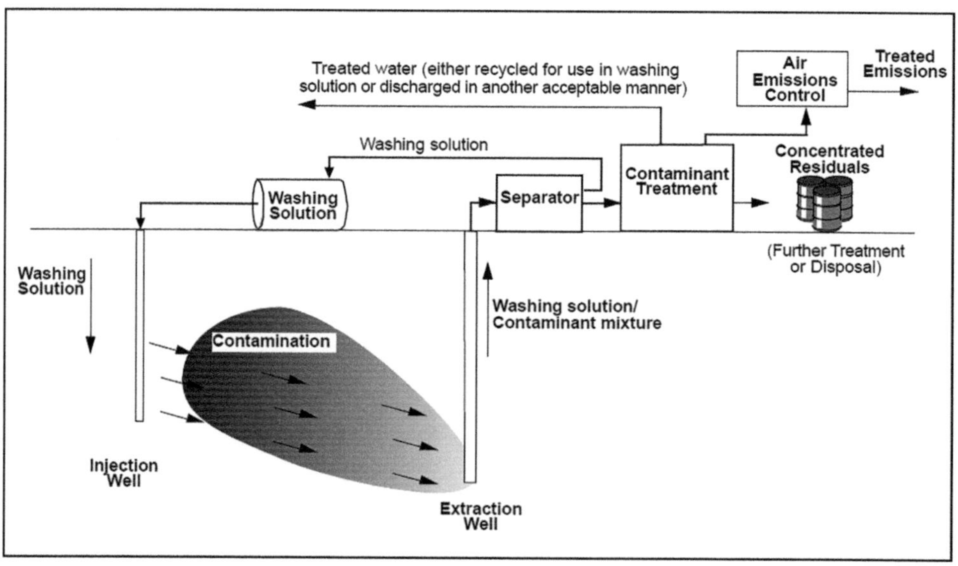

Figure 10.7 Schematic in situ soil flushing process using vertical wells (USEPA, 1996).

systems in the saturated zone. Unconventional techniques consist of primary, secondary, and tertiary recovery techniques (USEPA, 1993a). The in situ soil flushing involves drilling of injection wells and extraction wells into the ground where contamination has been located. The number of wells, their respective depth, and locations depend on the geologic factors and other technological considerations. Wells may be installed either vertically or horizontally. In addition to placing the wells, other equipment for treating the extracted wastewater must be transported to or built on the site (USEPA, 1996).

10.4.3 Cross-media transfer potential of in situ soil flushing

10.4.3.1 General

General cross-media transfer potentials during pre- and post-treatment activities are covered in Chapter 12. Soil flushing is different from most other technologies used to remediate contaminated soils; therefore, the concerns about cross-media contamination are fairly unique and are discussed in the following section.

10.4.3.2 Additional concerns for in situ soil flushing

Other concerns include various types of migration of contaminated fluids into surface or groundwater zones during the in situ flushing treatment. Types of potential releases are listed below:

- The primary waste stream generated is contaminated flushing fluid, which is recovered along with groundwater (Freeman and Harris, 1995). This fluid can cause cross-media contamination by migrating into uncontaminated groundwater zones, or if mismanaged, can be released into the surface environment.
- Treatment of the flushing fluid results in process sludges and residual solids, such as spent carbon and spent ion exchange resin, which may cause cross-media, transfer of contaminants if improperly managed and disposed of (Freeman and Harris, 1995).
- Residual flushing additives in the soil may be a concern and should be evaluated on a site-specific basis. These additives may require additional separation or treatment before disposal (Freeman and Harris, 1995).
- Bacterial fouling of infiltration, recovery systems and treatment units may be a problem, particularly if high iron concentrations are present in the groundwater or if biodegradable reagents are used.

10.4.4 Best management practices/options for controlling cross-media transfers during in situ soil flushing

General BMPs to prevent cross-media transfer of contaminants during pre- and post-treatment activities are covered in Chapter 12. BMP options to control specific cross-media transfer of contaminants during the in situ soil flushing treatment activities are furnished below:

- A thorough site characterization should be conducted to determine all leachable contaminants present.

- Depth to groundwater (including any seasonal variations) and presence of free product, if any, should be identified. All free products should generally be recovered before any treatment begins.
- During the construction of injection wells, care should be taken to properly:
  - o Anchor wells
  - o Install suitable screen meshes
  - o Protect the wellhead.
- The recovery system should have adequate capacity to collect injected fluids and groundwater, considering the maximum practical aquifer yield.
- If bacterial fouling of any part of the treatment system is a problem, the addition of compounds to control bacterial growth should be considered.

10.4.5 Waste characteristics that may increase the likelihood of cross-media contamination for in situ soil flushing technologies

The effectiveness of in situ soil flushing could be compromised under the following conditions and could cause undue cross-media contamination.

- Soils containing a high percentage of silt- and clay-sized particles typically are strongly adsorbed and are difficult to remove. These soils also tend to be less permeable (Freeman and Harris, 1995). The application of in situ soil flushing to soils with these characteristics may increase the need to use additives and reduce the efficiency of this contaminant removal method. Reduced contaminant removal may require the treatment of very large volumes of groundwater, which may increase the chances of cross-media contamination.
- Soils with low hydraulic conductivity (e.g., $K < 1.0 \times 10^{-5}$ cm/sec) will limit the ability of flushing fluids to percolate through the soil in a reasonable time frame (Freeman and Harris, 1995).
- Moisture content can affect the amount of flushing fluids required. Dry soils will require more flushing fluids initially to mobilize contaminants (Freeman and Harris, 1995).
- High humic content and high cation exchange capacity tend to reduce the contaminant removal efficiency of soil flushing (Freeman and Harris, 1995).
- Multiple factors, including high concentrations of fine sedimentary materials, inorganic precipitation, formation of stable emulsions, or excessive biological activity, can reduce the permeability required for successful treatment (USEPA, 1993a).

10.5 SOLIDIFICATION AND STABILIZATION

10.5.1 Definition and scope of solidification/stabilization

Solidification and stabilization (S/S) waste treatment processes involve the mixing of specialized additives or reagents with waste materials to reduce – physically or chemically – the solubility or mobility of contaminants in the environmental matrix (Freeman and Harris, 1995). S/S are closely related because both use chemical, physical, and thermal

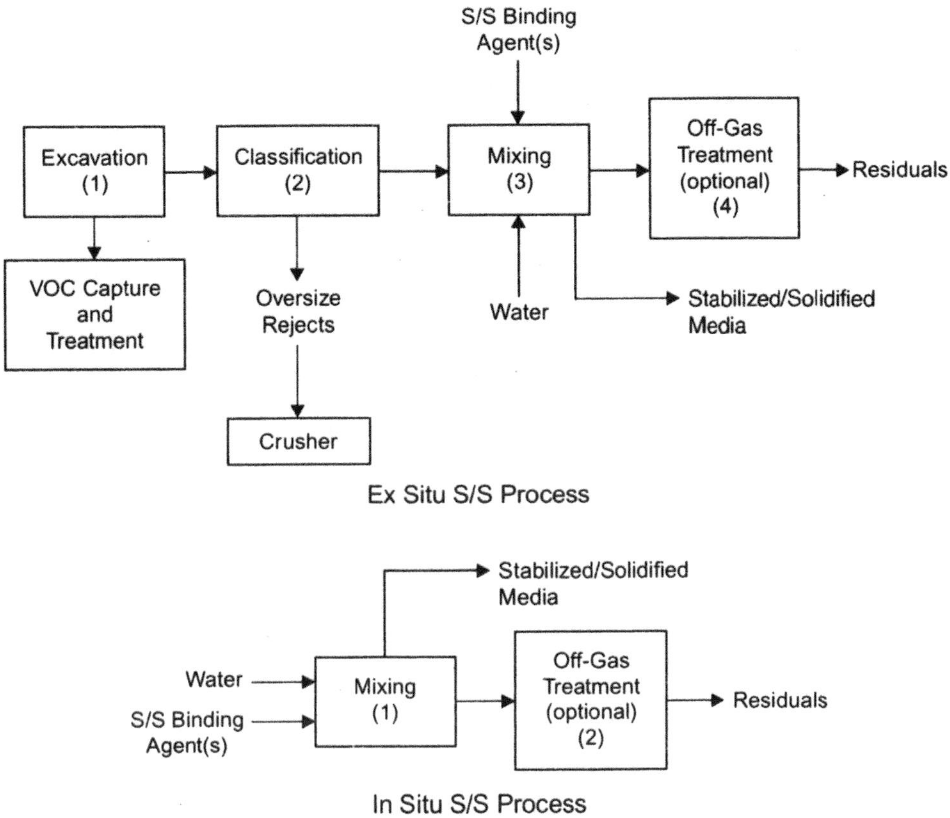

Figure 10.8 Shows a schematic of typical S/S processes (Dutta, 2002).

processes to detoxify a hazardous waste. However, they are distinct technologies that involve physical/chemical treatment processes (Dutta, 2002). Figure 10.8 shows a schematic of typical S/S processes.

10.5.1.1 Key features of solidification and stabilization

- Requires mixing of reagents, either on- or off-site.
- Immobilizes contaminants.
- Like other immobilization technologies, does not destroy inorganic waste, but may alter or change organic waste.
- Stabilization can be combined with encapsulation or other immobilization technology(ies).
- May increase total volume of materials that must be handled as waste.
- Wastes treated with S/S may be amenable to reuse following the treatment.
- Field application may involve installation of any or all of the following:
    - Auger-type drilling and mixing equipment for in situ applications
    - Dust collection systems

- o Volatile emission control systems
- o Bulk storage tanks.

10.5.2 Solidification and stabilization technology description

Solidification refers to processes that encapsulate the waste in a monolithic solid with structural integrity. The encapsulation may be that of compacted fine waste particles or of a large block or container of wastes. Solidification does not necessarily involve a chemical interaction between the waste and the solidifying reagents, but may mechanically bind the waste in the monolith. Contaminant migration is restricted by vastly decreasing the surface area exposed to leaching and/or by isolating the waste within an impervious capsule (USEPA, 1994b).

Stabilization refers to processes that reduce risk posed by a waste by converting the contaminants into a less soluble, mobile, or toxic form. The physical nature of the waste is not necessarily changed. Phosphates, sulfides, carbonates, etc. can be used as treatment reagents.

In many instances, stabilization is exclusive of solidification. Stabilized products should have low leaching characteristics. Many of the reagents used for S/S process are also used in other chemical treatment processes like dechlorination.

S/S systems can be used to treat contaminated soil or wastes in place or can be employed to treat excavated wastes externally for their subsequent disposal. S/S systems provide a relatively quick and lower-cost alternative to prevent exposure to contaminants, particularly metals. S/S systems have been selected or are being used in cleanups at over 250 highly contaminated (Superfund) sites across the US (USEPA, 2012b).

An innovative application of S/S technology involves reclamation of an old mine pit by using fly ash waste from a thermal power plant in Bokoshe, Oklahoma, USA (Dutta and Roper, 2011).

A summary of this unique S/S system involving beneficial use of fly ash for reclaiming an old mine pit is provided in the following section.

10.5.2.1 Solidification and stabilization of fly ash for beneficial use

This innovative approach used in Bokoshe, Oklahoma involves using the coal fly ash as waste from a nearby thermal power generation facility followed by stabilization of the fly ash to form a sealed lining of the old strip mine pits and conducting reclamation of these old abandoned mine pits.

The fly ash is brought in a specially designed hopper-bottom truck, which pulls up in a higher area to a unique unloading zone. The unloading area has a large capacity water tank and a truck bridge with an opening placed for the hopper-bottom truck to drop the fly ash. While the truck drops the fly ash through the opening of the hopper-bottom area, water from the tank is released simultaneously to mix the two. This prevents the emission of dust particles from the fly ash during unloading and it begins the process of making slurry at the same time. This slurry flows into a preliminary mixing area, which sits at about 40–45 feet higher than the water level of the mine pit. The slurry mix then overflows towards the strip pit and builds a sealed lining on the sides and at the bottom of the pit.

The operation is run in a natural setting with horses were grazing next to the mine reclamation pit. There is a natural pond, at a lower elevation from the mine pit, within 50 feet (15.2 m), which is maintaining a pH of 6.5–7.2 Standard Unit (SU) for the past two decades while this operation is ongoing. The mine pit under reclamation using the fly ash is highly alkaline and has a pH between 11.5 and 12.2 SU.

Figure 10.9 shows a picture (March 20, 2008) of the natural pond at a lower elevation and the mine pit used for fly ash reclamation at a higher elevation. The sky also looked quite clear in the picture. The fly ash unloading operation was ongoing at the site on that day using simultaneous mixing of water from stationary tanks and unloading of fly ash from the hopper-bottom trucks.

Several site visits were conducted during 2008–2011 to study and check on possible waste migration, other environmental impacts, and the balance of – social, environmental, economic, and scientific basis of this operation (Dutta and Roper, 2011). During a recent communication with the operator, Thumbs Up Ranch, LLC, confirmed the continuation of the integrity and sealing feature of the old mine pit by their observation of "no leaks" to the natural pond as evidenced by the high difference in color and pH of the water between the old mine pit and natural pond. They also indicated that they have stopped using the disposed saltwater from the local oilfields as intake water with high levels of total petroleum hydrocarbon (TPH) was contaminating the slurry and the pit water. They have replaced the intake saltwater with clean surface water from nearby areas. The operation is currently taking in about 30,000 tons of fly ash per year from the power plant and continuing to use that for reclaiming the old mine pits (Dutta, 2020).

Figure 10.9 Picture of the natural pond at a lower elevation from the fly ash disposal pit. (Photo: Subijoy Dutta, March 20, 2008.)

10.5.3 Cross-media transfer potential of solidification/stabilization

10.5.3.1 General

General cross-media transfer potentials during pre- and post-treatment activities for S/S are discussed in Chapter 12.

10.5.3.2 Additional concerns for solidification/stabilization

Other concerns include leaching and migration of contaminated fluids into surface or groundwater zones during the stabilization and solidification treatment process. Types of potential releases are listed below:

- Leaching of contaminants or excess reagents to groundwater from treated waste that is disposed on-site.
- Long-term degradation of the stabilized mass, creating conditions for the release of the solidified wastes, reagents, VOCs, and other contaminants from the treated waste.
- Leaching of contaminated fluids due to improper mix of the contaminated waste during the stabilization and solidification binding process (e.g., use of water contaminated with hydrocarbon or other organic substances in the mix).
- Extreme cold or rainfall may delay the binding period and/or release the contaminants if proper precautions are not taken.
- Dust or contaminants must be properly suppressed during the mixing. High wind or inclement weather may cause the release of contaminants into the air.

10.5.4 Best management practices/options for controlling cross-media transfers during stabilization/solidification

General BMPs to prevent cross-media transfer of contaminants during pre- and post-treatment activities are covered in Chapter 12. Specific BMP options to control cross-media transfer of contaminants during S/S pre-treatment and treatment activities are furnished below:

- Under dry and/or windy environmental conditions, both ex situ and in situ S/S processes are likely to generate fugitive dusts (Freeman and Harris, 1995). Refer to Chapter 12 for control mechanisms to reduce the cross-media contamination potential from fugitive dusts.
- Materials that are removed during prescreening activities should be disposed of properly.
- S/S processes can produce gases, including vapors that are potentially toxic, irritating, or noxious (Freeman and Harris, 1995). Vapor treatment systems should be used to the extent possible to control the movement of these vapors.
- If volatile organics are present, off-gas capturing and treatment systems should be designed according to recommendations provided in Chapter 13.
- Reagent delivery piping should be regularly checked to ensure tight fittings. This will reduce the chances of VOC releases.
- Wastes should be homogenized as much as practicable before processing. This can improve the efficiency of the stabilization activities, and may help to reduce spillage

and other problems related to the presence of irregular masses during the mixing process.
- Treated waste should be disposed of in a covered area and above the groundwater table.

10.5.5 Waste characteristics that may increase the chances of cross-media contamination for solidification and stabilization technologies

The effectiveness of S/S technologies could be compromised and undue cross-media contamination may be caused under certain conditions as given below:

- Physical mechanisms that can interfere with the S/S process include:
  - o Incomplete mixing due to the presence of high moisture or hydrocarbon/organic chemical content resulting in only partial wetting or coating of the waste particles with the stabilizing and binding agents and the aggregation of untreated waste into lumps (Dutta, 2020; Freeman and Harris, 1995).
  - o Disruption of the gel structure of the curing cement or pozzolanic mixture by hydrophilic organics in the soil (Freeman and Harris, 1995).
  - o Under mixing of dry or pasty wastes (Freeman and Harris, 1995).
- Chemical mechanisms that can interfere with the S/S process include:
  - o Chemical adsorption
  - o Precipitation
  - o Nucleation.
- Other factors that can interfere include:
  - o Precise tailoring of waste composition to the S/S process used (USEPA, 1994b).
  - o Waste containing oil and grease in moderate to high concentrations (Dutta, 2020).

However, it is possible to overcome some of these limitations with various technology-specific modifications and variations.

For more details on the various chemical interactions that can reduce the effectiveness of S/S treatments, please refer to technology-specific references provided at the end of this chapter (Freeman and Harris, 1995; USEPA, 1993b, 2012b).

10.6 EXCAVATION AND OFF-SITE DISPOSAL

10.6.1 Definition and scope of excavation and off-site disposal

When a site is remediated by excavation and off-site disposal, the contaminated material (typically a solid or semi-solid material such as soil or sludge) is excavated, then transported off-site for treatment and/or disposal.

10.6.1.1 Key features of excavation and off-site disposal

- Excavation or collection of contaminated soils, followed by piling or mixing of the soils.
- Containerization or temporary storage of the contaminated soils or solid media.

- Shipping of soils off-site for disposal.
- Field applications involve installation of a temporary canopy, liner, or other physical barriers as presented in Chapter 13 that minimizes movement of materials from the site by wind, water, or any other mechanism.

10.6.2 Excavation and off-site disposal technology description

Excavation and off-site disposal primarily involve equipment that is widely used in the construction or non-hazardous solid waste disposal industries, such as excavators, earthmovers or backhoes, dump trucks, and containers of various shapes, sizes, and materials. However, hazardous waste excavation and off-site disposal activities generally require that strict personal protection and safety measures are observed by everyone while they are engaged in these activities. These safety measures include special clothing, eye protection with glasses, hard hat, protective decontamination techniques, and equipment decontamination after the activity.

10.6.3 Cross-media transfer potential of excavation and off-site disposal

10.6.3.1 General

The general cross-media concerns are covered in Chapter 12. Among those, the ones that refer to excavation and construction activities are especially relevant to this technology. Table 12.1 presents information on the contribution of remedial activities, including excavation, materials handling, and transportation – that are of concern to all remedial technologies including this option.

10.6.3.2 Additional concerns for excavation and off-site disposal

Concerns for excavation and off-site disposal center around the potential for cross-media transfer during materials handling and transportation activities. Careful attention should therefore be paid to information presented in Chapters 12 and 13, particularly as it relates to the handling and transportation of contaminated wastes.

10.6.4 Best management options for controlling the cross-media transfer potential of excavation and off-site disposal

General BMPs to prevent potential cross-media transfer of contaminants during excavation and off-site disposal activities are covered in Chapters 12 and 13. The most applicable BMP options for this technology, as presented therein, include the following:

- Limiting entry to the active site to avoid unnecessary exposure and related transfer of contaminants, especially during site preparation and staging.
- Avoiding entrance to the contaminated area. In unavoidable circumstances, build a temporary decontamination area, which could be later used during cleanup activities.

Any above-ground and underground source of contaminants should be identified and located before starting excavation of the contaminated area.

- Fugitive dust emissions should be controlled during excavation by spraying water or other materials to keep the ground moist or covered. During wet weather or rainfall, no water spraying would be needed. See Chapter 13 for more information on materials that can be used to control fugitive dust emissions.
- During the transportation of contaminated soils or solid media, covers or liners should be used to prevent dust and VOC emissions. These temporary covers on trucks or other hauling equipment should be installed with care to minimize the possibilities for the waste to come into contact with high winds during transport.
- Any offsite runoff should be prevented from entering and mixing with on-site contaminated media by building earthen berms or adopting similar other measures, as outlined in Table 13.4.
- Provisions should be made to capture on-site surface water runoff by diverting it to a controlled depression area or lined pit.
- Covers, and if necessary, liners, should be used at all times when contaminated materials are being stored. Covers should be used on trucks that are moving materials around and from the site. See Chapter 13 for details on covers and liners that should be considered for use during excavation, storage, and transportation.

10.6.5 Waste characteristics that may increase the likelihood of cross-media contamination for excavation and off-site disposal technologies

Under the following conditions, the effectiveness of excavation and off-site disposal could be compromised and could cause undue cross-media contamination.

- This option may not be very effective when high volumes of soils are to be disposed of.
- In the case of highly explosive materials, a simple excavation and disposal may not provide the safest method of handling the waste.

REFERENCES

Campbell, B. E. (1995) *Geosafe Corporation, Comments on BMP Workshop Summary-Other Physical/Chemical Technologies. Personal Communication to Subijoy Dutta.* USEPA.

Dev, H. (1995) *IIT Research Institute, Letter Report on RF In Situ Heating and Soil Decontamination Process. Personal Communication to Subijoy Dutta.* USEPA.

Dutta, S. (2002) *Environmental Treatment Technologies for Hazardous and Medical Wastes, Remedial Scope and Efficacy.* Tata McGraw Hill Publishing Company, New Delhi.

Dutta, S. (2020) S&M Engineering, LLC (Personal communication with Thumbsup Ranch, LLC August 28, 2020), Crofton, Maryland, USA.

Dutta, S., E. W. Roper. (2011) Presentation of "An Innovative Approach for Fly ash Management" at AES Power Plant, March 25, Shady Point, Oklahoma.

Freeman, H. M., and E. F. Harris. (1995) *Hazardous Waste Remediation: Innovative Treatment Technologies.* Technomic Publishing Co., Inc., Lancaster, PA.

USEPA. (1990) *Engineering Bulletin-Chemical Dehalogenation Treatment: APEG Treatment.* EPA/540/2–90/015, Office of Research and Development, Cincinnati, OH, September.

USEPA. (1991) *Engineering Bulletin-In Situ Soil Flushing.* Office of Research and Development, Cincinnati, OH. EPA/540/2-91/021, October.

USEPA. (1992) *Guide for Conducting Treatability Studies Under CERCLA Solvent Extraction, Interim Guidance (and Quick Reference Fact Sheet).* Office of Emergency and Remedial Response. EPA/540/R-92/016A and B, August.

USEPA. (1993a) *Innovative Site Remediation Technology: Soil Washing/Soil Flushing.* Vol. 3. EPA/542/B-93-012, November.

USEPA. (1993b) *Technical Resource Document-Solidification/Stabilization and its Application to Waste Materials.* Office of Research and Development, Washington, DC. EPA/530/R-93/012, June.

USEPA. (1994a) *Engineering Bulletin-In Situ Vitrification Treatment.* Office of Research and Development, Cincinnati, OH. EPA/540/S-94/504, October.

USEPA. (1994b) *Innovative Site Remediation Technology, Solidification/Stabilization.* W.C. Anderson, ed. Vol. 4. Office of Solid Waste and Emergency Response. EPA/542/B-94/001.

USEPA. (1995a) *SITE Technology Capsule: IITRI Radio Frequency Heating Technology.* EPA/540/R-94/527a, March.

USEPA. (1995b) *Innovative Technology Evaluation Report, Radio Frequency Heating.* KAI Technologies. EPA/540/R-94/528, April.

USEPA. (1995c) *In Situ Remediation Technology Status Report: Surfactant Enhancements.* Office of Solid Waste and Emergency Response. EPA-542-K-94-003.

USEPA. (1996) *A Citizen's Guide to In situ Soil Flushing.* Washington, DC. EPA 542-F-96-006, April.

USEPA. (2012a) *A Citizen's Guide to In situ Thermal Treatment.* Washington, DC. EPA 542-F-12-013, September.

USEPA. (2012b) *A Citizen's Guide to Solidification and Stabilization.* Washington, DC. EPA 542-F-12-019, September.

Chapter 11

Case studies of treatment technologies

This chapter provides information on the field applications of environmental treatment technologies for remediation purposes. It focuses on the field applications of various remediation treatment technologies and the use of relevant best management practices (BMPs) for controlling cross-media transfer/migration of pollutants.

11.1 SCOPE OF CASE STUDIES OF TREATMENT TECHNOLOGIES

The "case studies of treatment technologies" covers in detail the applications of various treatment technologies used for remediating contaminated sites. Most of these applications will fall under one of the seven categories of remedial treatment technologies covered in the previous chapters. It also includes some novel treatment technologies that are currently being applied to the treatment of contaminated soil and other media. Keeping in mind the ultimate goal of total protection of the environment, the cross-media transfer aspect and problems encountered in the field applications are primarily the focus of these case studies (Dutta, 2002; USEPA, 1997, 2000).

The objectives of the case studies presented in this chapter are to:

- highlight the treatment technology applications observed at each site,
- elaborate on how certain BMPs were used at different sites, and
- compare the recommended BMPs with the current practices at various sites.

The information contained in this chapter has been developed by examining detailed site work plans, soil treatment reports, and by reviewing reports on remediation activities at various sites. Summary information from various field studies conducted by EPA (USEPA, 1997, 2000) and personal communications (Dutta, 1996, 2000) have been compiled and presented here. In total, information on remedial treatment applications was obtained from various sites, mostly in the US. The sites in the US are located in the states of Colorado, Connecticut, Maine, Maryland, Minnesota, and Virginia, which primarily cover the following types of soil treatment technologies: containment, soil washing/soil leaching, soil-vapor extraction, thermal treatment, bioremediation, phytoremediation, and chemical-based stabilization.

Table 11.1 summarizes key characteristics of the sites studied. Most of the case studies are based on records from EPA publications (USEPA, 1997, 2000).

Sections 11.1–11.8 provide details on the field use of remedial treatment technologies applied at each site as well as focus on the BMPs applied at these sites. A brief description of site

DOI: 10.1201/9781003004066-11

Table 11.1 Key characteristics of case study sites (Dutta, 2002; USEPA, 1997)

	Location	*Type of contamination*	*Area/depth of contamination*	*Treatment technology*	*Volume treated*	*Duration/status of site remediation*
1.	Army Ammunition Plant, Minnesota	Metals in soil	16 acres/2 feet below surface	Soil washing/soil leaching	24,748 tons	From 1993 to 1995 full scale
2.	Petroleum Refinery, Minnesota	Petroleum-derived hydrocarbons	Two lagoons 0.5 acre/to a max depth of 10 feet	In situ soil stabilization	18,000 cubic yards	Conducted in 1995. full scale
3.	Closed Battery Manufacturing Facility, Virginia	lead	Plant site of 4.5 acres/3 feet deep	Ex situ soil stabilization	25,578 cubic yards	Conducted in 1993 full scale
4.	Ammunition Testing/Disposal Site, Connecticut	Metals (primarily lead)	Scattered distribution of contaminated areas over 435 acres/surface	Size separation/soil washing	1,000 cubic yards	Pilot study completed in 1995
5.	Closed Electronics Component Manufacturing Facility, Maine	Spent organic solvents	below a building, 3 acre site	Soil vapor extraction	Estimated 40,-000–50,000 cubic yards	Conducted during 1995–1997 full scale
6.	Closed Chromium Manufacturing Facility, Maryland	Chromium	15–20 acres; depth 80 feet or bedrock	Site containment	Estimated 1–2 million cubic yards	1991–1999 – Phase I completed
7.	DoD Manufacturing Facility, Virginia	Organic compounds	300 foot long drainage area; 4–5 feet wide	Ex situ bioremediation	500 cubic yards	Pilot completed in 1996
8.	Amcor Precast, Ogden, Utah	Groundwater (GW) and soil	Groundwater 1.5–3.5 m	Air sparging and soil vapor extraction	GW – 5,100 m^3 (1.35 million gallons), soil – 5,356 m^3	March 1992–September 1993 full-scale cleanup
9.	Arlington Blending, Tennessee	Pesticides and metals in soil	Excavated soil – 10,000 tons	Thermal desorption rotary dryer (580°F–750°F)	41,431 tons	January–June 1996 full-scale cleanup –
10.	Avco Lycoming, Williamsport Pennsylvania	Groundwater (GW) and soil-chlorinated solvent	Groundwater – at about 25 ft depth covering 2 acres	In situ bioremediation; anaerobic reductive dechlorination	Estimated at > 1.5 million gallons	February 1997–June 1997 pilot and full-scale cleanup

remediation activities at the site, an in-depth description of the BMPs used, and some views and discussion on the specific case study have been presented in each of these eight sections. In the last four case studies (Sections 11.9–11.12), the emphasis is primarily on factors, such as cost, new and innovative technologies, and the efficacies of these technologies in terms of site remediation. At the end, under Section 11.13, a comparison of selected case studies with respect to their specific site application characteristics has been presented. It compares the remedial treatment technologies and the respective BMPs used versus the recommended BMPs, outlined in Chapters 4–10 (Dutta, 2002).

11.2 SOIL WASHING/SOIL LEACHING TO TREAT METALS CONTAMINATED SOIL AT AN ARMY AMMUNITION PLANT IN MINNESOTA (SITE 1)

11.2.1 Description of site remediation activities

In an Army ammunition plant in Minnesota, during 1993–1995, metals contaminated soil was excavated from a 16-acre area and treated by soil washing/soil leaching at the site. The treatment goals were to reduce the concentration of lead to a regulatory level of 300 mg/kg and the concentration of several other heavy metals to the background levels. The site was previously used as an open burning area for scrap primers, fuses, and explosives related to small caliber arms and rifle grenades. The metals contamination was generally limited to the uppermost three feet (ft) across the site. The site was not contaminated with VOCs, SVOCs, or cyanide. The groundwater was not impacted by soil contamination. However, twenty disposal areas were identified at this site, which were also excavated and remediated. The materials found in the disposal areas included ordnance, high explosive items, cast-iron pots, crushed drums, characterized chemical substances, miscellaneous scrap metal, wood, concrete, and glass debris. All materials excavated from the disposal areas were identified and sorted for proper treatment or disposal off-site.

The site remediation was performed according to closure plans approved under a RCRA Hazardous Waste treatment storage and disposal (TSD) Permit and the Federal Facility Agreement (FFA) prepared for the Installation Restoration Program (IRP) of the Army ammunition plant. The site closure activities included site preparation, disposal area investigations, and excavation. It included metals-contaminated soil excavation and treatment, ordnance clearance, hazardous and non-hazardous waste management, and site restoration to allow unrestricted future use of the site.

The summary results from this remediation treatment are listed below:

- Approximately, 24,748 tons of contaminated soil were excavated at this site.
- Approximately 12,797 tons of soil were successfully treated and backfilled as clean soil.
- About 7,125 tons of soil were treated to become non-hazardous and stored temporarily at the site.
- About 246 tons of metal concentrate was recovered by the soil washing/soil leaching activities.
- Approximately 4,555 tons of soil, characterized as hazardous and non-treatable on-site, were recycled off-site at a smelter.

11.2.2 BMPs used to prevent the cross-media transfer of pollutants

Because the remediated site was located in a large federal facility (2,300 acres) where several private companies operated as facility tenants, there was the potential hazard for people in the area being exposed to lead and other toxic metals from fugitive dust emissions during site excavation, soil transportation, and soil treatment. Additional threat was posed by the cross-media transfer potential of pollutants in uncontrolled surface water run-on to and run-off from the site. Yet another concern was that the spillage and/or improper disposal of soil treatment residuals might result in groundwater contamination. To address these cross-media transfer concerns, various BMPs were used at this site during site remediation. The details on the field application of 12 BMPs, as introduced in different remedial stages, are provided below:

A. Site preparation and staging
   a. Delineating the actual limits of soil contamination:

      Additional soil borings and field analysis of total lead concentrations were performed as one of the earliest activities of the site remediation process in 1993 to delineate further the previously estimated limits of soil contamination. A total of 326 soil borings were made to collect soil samples at 6- and 12-inch depths every 25 feet along the estimated limits. X-ray fluorescence (XRF) analysis was used to determine the soil quantity in the field. If samples from both depths at a given soil boring were shown to have XRF lead concentrations of less than 100 ppm, the location was considered non-contaminated. Additional XRF samples were then collected in an attempt to delineate the boundary of contamination within five feet of the actual zone of contamination (which also represented the horizontal limits of excavation). These activities limited the total quantities of soil excavated and treated during site remediation.
   b. Establishing an exclusion zone and a decontamination pad to control all traffic:

      After verifying the boundaries of contamination with laboratory analysis of soil samples, the exclusion zone was established by providing a buffer zone outside the finalized limits of contamination. Access to the exclusion zone was limited to the personnel wearing appropriate personnel protective equipment and qualified to perform only the site remediation activities approved by the site closure plan. After establishing the exclusion zone, a decontamination pad was constructed for vehicle, equipment, and personnel decontamination. All traffic into and out of the exclusion zone went through the decontamination pad. The decontamination pad was a 6-inch thick asphalt pad underlain with a gravel subbase. The pad was constructed with a 3-inch high asphalt berm around the perimeter to control run-off and run-on. All water used in the decontamination pad was collected in a 100-gallon sump. Both the pad and sump were coated with sealant to help prevent any seepage of contaminated water. Two 8-foot high wind walls were constructed to minimize decontamination spray from migrating off the pad. A fully enclosed personnel decontamination area was also constructed on the decontamination pad. The construction of a decontamination pad with all these features enabled the facility to manage effectively any release of pollutants taking place during the excavation and/or transportation of contaminated soil at the site.

c. Refurbishing the existing oil treatment pad/area:

An existing concrete pad located near the site was utilized for soil washing and soil leaching operations. The concrete pad was 6–8 inch thick and had masonry block walls. However, the pad was used only after preparing it for soil washing/soil leaching equipment. The soil feed section of the pad used for stockpiling contaminated soil was completely resurfaced with adequate slope for drainage of water to a sump. Several large, deep cracks existing in other portions of the treatment pad were also sealed to ensure further containment of pollutants released from the soil or from the chemicals used on the pad during the soil treatment process. A new drive-over curb was constructed to prevent contaminated water generated in the soil washing area from entering the treated soil storage area. These preparatory efforts enabled the facility to contain effectively any release of pollutants taking place during soil treatment on-site.

d. Conducting an in-depth characterization of new disposal areas located on-site:

Before any soil excavation activities, certified explosives and unexploded ordnance contractors performed a visual ordnance survey within the exclusion zone. Based on the quantity of ordnance identified on the surface of the site, it was decided to take up additional investigative work on-site. A magnetometer survey performed across the site identified 24 soil anomalies. Investigative test trenching at the locations of these anomalies and other suspect areas identified 11 disposal areas. Nine more disposal areas were identified following excavation of metals contaminated soils through a review of historical aerial photographs or verification test trenching.

e. Removing and sorting the debris found in disposal areas:

The disposal areas were excavated before starting the excavation of metals contaminated soils. As a result, the metals contaminated soils were segregated from the debris buried in the disposal areas. The materials found in the disposal areas included high explosive items with approximately ten pounds of explosives. These materials were untreatable by soil washing/soil leaching. These items were detonated at two locations on-site. All other debris were sorted and sent off-site for recovery or disposal.

f. Installing an air quality monitoring system at the site:

The health and safety program at this site included the monitoring of specific air pollutants during the entire period of remediation at disposal areas, metals contaminated soil excavation, and on-site treatment of soil. For example, the investigative test trenching and disposal area excavation were conducted without encountering hazards from pollutants, i.e. action-level exceedances for airborne lead, combustible gases, oxygen, total dust, and hydrogen cyanide. The maximum level of total dust was 0.070 mg/cu.m. (action level = 0.30 mg/cu.m.). The maximum airborne lead was 0.009 mg/cu.m. (action level = 0.03 mg/cu.m.). The air quality monitoring was performed along with the monitoring of other airborne pollutants, such as noise and heat stress in the work areas.

B. Pre-treatment

a. Building an earthen berm for controlling surface water run-on to and run-off from the site:

A soil berm approximately three feet wide and 2 feet high was constructed immediately outside the exclusion zone. The berm was 1,330 linear feet, and it surrounded the site completely. It was constructed to contain contaminated stormwater precipitating on the site as well as to prevent surface water from flowing into the exclusion zone.

b. Arranging for dust suppression during site remediation:

In preparation for encountering adverse working conditions, the site was provided with adequate dust control measures. These measures included a 1.5-inch PVC water line constructed along the site access road with sprinklers tapped into the line every 25 feet. The line was used to wet the road, thereby minimizing fugitive dust along the road during the transportation of contaminated soil to the treatment pad. A 3/4-inch black PVC water line was also run along the boundary of the exclusion zone with valves located every 100 feet so that the entire site would have access to water.

c. Covering haul trucks and excavated soil stockpiles:

The excavated surface soils were hauled from the site to the treatment pad in dump trucks covered with a tarp to prevent spillage and dust emission. The transported soil was stored in the stockpile area of the treatment pad. Additional quantities of excavated soil were stockpiled within the exclusion zone near the decontamination pad. Stockpiles were maintained only to provide adequate quantities for continuing on-site treatment operations. They were covered with reinforced plastic sheets to prevent fugitive dust emission and rainwater infiltration. The same contaminated soil handling procedures were used for excavation and transport of soils from the disposal area, as well.

C. Soil treatment

a. Reusing stormwater as process makeup water:

The stormwater run-off from the impervious soil treatment pad was collected in a 20,000-gallon holding tank. Stormwater was also collected in sumps provided in the soil storage pad. To whatever extent possible, water in the holding tank was used as makeup water in the soil treatment process, thereby eliminating the need for additional treatment of the contaminated stormwater.

b. Recycling process wastewater in soil treatment:

Process water used in soil washing/soil leaching was normally recycled throughout the project. Recycling of the process water minimized the generation of process wastewater.

D. Post-treatment

c. Reusing treated/clean soil in restoring the site:

The treated/clean soil was used to backfill the site. The site was regraded, provided with clean top soil and revegetated for better soil erosion control. By using treated soil in restoring the site, it was possible to prevent the disposal of large quantities of excavated soil (with residual contamination) off-site.

11.2.3 Views and discussion

In addition to the practices listed above, efforts were made to minimize the need to dispose of hazardous soil treatment residuals off-site. For example, during the first year of site remediation (1993), batches of soil that met treatment standards for any of the heavy metals of concern were hauled back to the contaminated soil feed pad for reprocessing. These batches were placed in stockpiles where some mixing of these with other soils took place. The batches failing to meet treatment standards during the next two years (1994 and 1995) were first assessed to rule out their possibility of being managed as non-hazardous waste. When the TCLP results indicated that the soil from a particular batch was non-hazardous, then the batch was transported to a separate soil storage pad. The other batches failing to meet treatment standards were disposed of off-site.

The residuals from the metal contaminated soil treatment also included metal concentrates generated in the density separation stage and electrowinning unit of the soil washing/soil leaching process. Metal concentrates were accumulated in drums and manifested to a smelter for metal recovery. Any treated soil found to be hazardous was sent off-site for metal recovery.

Following completion of soil excavation and treatment activities, the decontamination pad built near the site was excavated and treated. Wipe samples were collected on the surface of the decontamination pad before the commencement of any work and after completion of all work on the pad. The material excavated from the pad was then disposed of as non-hazardous waste.

11.3 IN SITU CHEMICAL-BASED STABILIZATION OF PETROLEUM-CONTAMINATED SOIL AT A REFINERY IN MINNESOTA (SITE 2)

11.3.1 Description of site remediation activities

In 1995, two previously used biological treatment facilities (aeration lagoons), part of a refinery's wastewater treatment plant in Minnesota, were closed after implementing in situ chemical-based stabilization for approximately 18,000 cubic yards of petroleum-contaminated soil at the facilities. Both the lagoons had occupied the eastern portion of an area of fill between a bedrock escarpment and the Mississippi River. The topography of the fill was relatively flat, rising slightly to the north. The original lagoon bottoms were approximately 15–20 feet below the ground surface and lagoon elevations were at the same level as the river.

Both lagoons were identified in 1990 to be hazardous waste management facilities due to the occasional presence of benzene in concentrations exceeding the toxicity characteristic limit in the wastewater entering the first of two lagoons, which were in a row. A RCRA Part B permit application to continue operation of the lagoons was then submitted to EPA and the Minnesota Pollution Control Agency (MPCA). However, it was decided to close the existing facilities permanently after the initiation of a groundwater monitoring program in the vicinity of the lagoons and investigation of the pollutants in the subsoil in both of these lagoons. A partial stabilization of the subsoil in the lagoons was then performed by mixing the subsoil up to a depth of six feet with cement using a track-mounted backhoe equipped with injector tines. Based on the results of subsoil investigation conducted after these preliminary soil remediation efforts, it was decided to perform additional stabilization of subsoil in the lagoons. The refinery submitted a revised closure plan for the two lagoons in 1994 and finally amended the revised closure plan early in 1995.

Pursuant to the closure plan approved under RCRA, the subsoil in both lagoons was stabilized in situ by utilizing a mixture of portland cement and fly ash (20%:80%) as a stabilizing agent or grout to reduce the potential for leaching the VOCs, SVOCs, and metals from the stabilized materials. The stabilization matrix is also expected to reduce the generation of leachate and prevent the long-term release of organic compounds to the groundwater from the soil underlying the stabilized materials in the lagoons. In situ grouting of subsoil was performed by drilling into the subsoil with a 10-foot diameter auger and mixing the soil in the borehole with grout as the auger was retrieved. At each drilling and grouting location, a column of stabilized material of low permeability was created which contained and immobilized contaminants within the soil. Before the implementation of this soil treatment process, the lagoons were emptied of oily sludge and wastewater.

11.3.2 BMPs used to prevent the cross-media transfer of pollutants

There was a potential for surface water pollution in the initial stages of remediation at this site (e.g., removal of large quantities of oily sludge and wastewater from the lagoons before commencing soil treatment) due to the proximity of a river to the site. Moreover, soil contamination at the site had adversely impacted groundwater in the area and the quality of groundwater had to be monitored for several years to determine if soil treatment had been effective. The details on the field application of the following five BMPs, as introduced in different remedial stages, are provided below:

A. Site preparation
   a. Removing and containing liquid wastes at the site:
      Oily sludge and wastewater removed from the lagoons were collected in portable tanks before carrying out soil remediation. Earthen berms (1-foot high and 3-feet wide) surrounded each portable CAMU Padtank. A poly tarp was placed on the ground and berms before the portable tanks were set in place. The earthen berms and poly tarp were provided to take care of the spills of the process treatment residuals occurring near the tanks. The tanks were placed outside the remediation site, near a sharply sloping river embankment.

B. Pretreatment
   a. Preventing surface water run-on to the site:
      The lagoons were bound by man-made dikes on three sides and a 20 feet natural escarpment on the fourth side. The higher ground beyond the escarpment was occupied by the refinery, which was on a well-paved, graded, and diked area. Because the soil stabilization area was in a floodplain of the Mississippi River, the top of the west dike facing the river was elevated by 3 feet as an additional precaution against flooding during springtime.

C. Soil treatment
   a. Treating accumulated stormwater on-site before disposal:
      There was ample freeboard (13 feet) to prevent run-off from the soil stabilization area. The accumulated stormwater was pumped for treatment to an adjacent wastewater treatment plant. The pumps and wastewater treatment plants had adequate capacity for this purpose.
   b. Conducting additional soil tests to determine the effectiveness of soil stabilization:
      Confirmatory soil sampling was conducted to obtain additional information on the depth of soil stabilization and soil quality. The soil samples were examined visually and for odor and screened for organic vapors using the jar headspace method. The stabilized materials and unstabilized subsoils below were then analyzed to determine the leachability of organics and metals present in the samples.

D. Post-treatment
   a. Monitoring groundwater to determine the effectiveness of soil stabilization:
      If the stabilization matrix became less effective over time, the VOCs and SVOCs released before site remediation would be observed at increasing concentrations in the groundwater near the site. A groundwater demonstration monitoring was started after completing soil stabilization at this site. This activity included regular monitoring of groundwater sampled at wells up- and down -gradient of the lagoons, as well as in aquifers near the river. The closure plan for the site called for groundwater monitoring every month during the first year; every quarter in the second and third year; and semiannually in the fourth and fifth year after the site remediation was completed.

11.3.3 Views and discussion

This site provided many opportunities for better addressing some of the major cross-media transfer concerns that existed. For example, the berms around the portable tanks used to collect and remove oily sludge and wastewater from the lagoons were not high enough to protect against a major spill of the liquid waste being removed from the site. The flexible hose from the lagoons to a tank became loose during the remedial activity at this site and flopped out of the containment. Nearly 2,000 gallons of oily sludge spilled into the adjacent river before the spill could be stopped. If a site-specific emergency response plan was in place and had been implemented, a major spill might have been prevented in this case. After the incident, earthen berms and poly tarp were provided, as mentioned earlier in this case study, to contain spills as one of the BMPs.

The field screening and analysis of soil samples at both of these lagoons had shown the presence of petroleum hydrocarbons in relatively high concentrations; including VOCs like benzene and xylene and SVOCs like 2-methylnapthalene and phenanthrene. Under these site conditions, fugitive emission of VOCs and other gases need to be controlled by monitoring the volatilization of organic compounds during soil treatment. The soil stabilization process injected liquid slurry by means of an auger, which helped avoid creating a large area of soil-air contact and the volatilization of organic compounds present in the soil. An air injection technology for the stabilizing agent might have resulted in significant air emissions of volatile organics. Under these conditions, it would have been appropriate to provide for the collection and treatment of VOCs released from the site. Air quality monitoring could have also addressed the release of VOCs from the stabilized soil.

Although some arrangements were made to prevent surface water run-on to the site (see "B" under Section 11.3.2), it is important that the likelihood of the worst run-on scenario be taken into account in sites located within the floodplain, because of a possible breach of the dike under a heavy flooding condition may result in a wash out of the lagoon contents and contaminated soils into the river.

11.4 STABILIZATION AND DISPOSAL OF LEAD-CONTAMINATED SOILS AT A CLOSED BATTERY MANUFACTURING FACILITY IN VIRGINIA (SITE 3)

11.4.1 Description of site remediation activities

In 1993, a phosphate-based stabilization technique was successfully performed on approximately 25,578 cubic yards of lead-contaminated soil and debris found in 11 acres of land at a closed battery manufacturing facility site in Virginia. The site was contaminated with lead concentrations in soil exceeding 100,000 mg/kg. The contaminated soil exhibited toxicity characteristics by virtue of the toxicity characteristics leachate procedure (TCLP), with lead concentrations in the leachate as high as 345 mg/L.

The phosphate-based stabilization process employed at the site significantly decreased the leachability of lead in soil, determined by TCLP, to levels well below the regulatory threshold of 5 mg/L. The stabilization process yielded a disposable material at greatly reduced costs. Soil remediation included excavating and mixing lead-contaminated soil with triple superphosphate (TSP), water and magnesium oxide (MgO) in a pug mill. Lead phosphate, which is one of the least soluble and most stable forms of lead in the soil environment, was formed. The action level for treating most of the lead-contaminated soils found at the site was kept at 1,000 mg/kg. Lower action levels were used for lead-contaminated soils found in the drainage ditch and sedimentation basin used for stormwater control at the site (400 mg/kg), as well as a previously used acid pond on-site (250 mg/kg).

11.4.2 BMPs used to prevent the cross-media transfer of pollutants

During the preparation of this site for remedial action, it was necessary to prevent any removal of contaminated soil from the site without appropriate treatment. Fugitive emission of dust containing lead and the "carry-over" of lead by stormwater run-off were the main concerns during pre-treatment and soil treatment at this site. The main concern during post-treatment was the fact that action levels (1,000 mg/kg) for most of the site might have resulted in concentrations of lead in soil remaining at the site being substantially higher than background levels. The details on the field application of the following six BMPs, as introduced in different remedial stages, are provided below:

A. Site preparation and staging
   a. Establishing site control zones to limit access to site:

      The site fence was relocated before any soil disturbance to provide security for the exclusion zone, which contained all soils with lead concentrations above 1,000 mg/kg. Access to the exclusion zone was permitted only through the contamination reduction zone, where a pad was constructed to serve as a transfer area for all personnel and equipment. All decontamination water was collected and discharged to two 500-gallon settling tanks. The tanks were placed in a row and they facilitated the settling of sediments from the decontamination water. Lead and 10 micron (μm) particulate filters were installed on the discharge pipe of the second settling tank to ensure adequate removal of suspended solids. The recycled water was then re-used for decontamination purposes. If water reuse was not possible, the decontamination water was disposed of off-site. The filters were periodically replaced.
   b. Performing site clearing and grubbing with limited off-site removal of pollutants:

      After the health, safety, and environmental monitoring controls were in place, but before beginning the excavation activities, designated areas on the site were cleared and grubbed to minimize off-site transfer of lead-contaminated soils without treatment. All the grubbed materials were decontaminated and then disposed of off-site. Other building structures were dismantled on-site and decontaminated with a high-pressure water spray before the structures were disposed or recycled. This action also minimized the off-site transfer of lead-contaminated soils without treatment.

B. Pretreatment

a. Installing an air monitoring system and on-site weather station:

A Perimeter Air Sampling Program (PASP) was implemented to document the impact of on-site remedial action on the local ambient air quality. The critical concentrations to which ambient air sampling results were compared were 150 g/m^3 for particulates less than 10 µm in diameter (PM-10) and the 1.5 µg/m^3 for lead. Air monitoring was also used to ensure the effectiveness of dust control at the site. There was only one exceedance of the critical air concentrations during the entire site remediation period. It was due to fugitive dust caused by dry soil conditions and heavy truck traffic. Corrective action was taken immediately by spraying additional water on excavated areas and stockpiled materials to suppress fugitive dust emission. An on-site weather station was used to control soil handling and excavation activities during high winds.

b. Constructing a drainage ditch and sedimentation basin for stormwater run-off control: A sedimentation basin was constructed to provide for the collection and recycling of contaminated runoff from the site and the settling of any contaminated sediment from stormwater run-off during the entire site remediation period. The basin was designed to retain runoff by providing two hours of retention time for all run-off generated by a 25-year, 24-hour storm event. The retention time was adequate to settle particles as small as 10 µm in diameter. Throughout the remedial process, collected stormwater was recycled from the sedimentation basin to the pug mill for inclusion within the solidification/stabilization process. However, this volume did not constitute all of the make-up water for pug-mill operations. An unspecified volume was transported from off-site when the water level in the sedimentation basin was too low. Monthly samples of the water from the stormwater retention basin and quarterly samples of the surface water, which intermittently appeared in the drainage ditch, were collected and analyzed. The samples were used to confirm that off-site contamination of surface water had been prevented. Quarterly samples of the sediments were also taken in the drainage ditch upstream and downstream of the sedimentation basin. No significant change was found in the lead concentrations of the sediments obtained downstream of the sedimentation basin. As expected, lead concentrations increased in the sediments obtained throughout the period of site remediation from the drainage ditch upstream of the sedimentation basin.

C. Soil treatment

a. Varying the treatment process at different areas of the site most effectively:

After the construction of the sedimentation basin first, the excavated soil was separated into two storage areas: one area containing contaminated soils and the other area containing soils below the performance standard for lead of 1,000 mg/kg for this site. The soil was then excavated from the remaining site, sampled and stored accordingly. As noted earlier, the performance standard for soils found in the previously used acid pond was kept much lower at 250 mg/kg, and a performance standard of 400 mg/kg was used for soils found in the drainage ditch and sedimentation basin after completing soil excavation and treatment in the remaining site. By segregating the soil with lead above performance standards throughout the project, it was possible to minimize the total quantity of soil to be stabilized and transferred off-site for disposal. Before dismantling of the pug-mill, the areas used for stockpiling contaminated soil were the last to be excavated. With dismantling of the pug-mill, the surrounding areas used for staging, treatment, and loading of treated soils were resampled, excavated, and solidified or stabilized in roll-on containers, as required. Soils from the drainage ditch and the sedimentation basin were finally excavated and disposed of. However, these soils were stabilized in situ to avoid a transfer of contaminated water from the wet soils excavated at these site locations.

D. Post-treatment
   a. Backfilling the excavated areas and site vegetation:
      After receiving analytical results confirming that performance standards were achieved for all soil and sediment samples obtained from the site after treatment, the excavated areas were backfilled with site soil containing lead below 1,000 mg/kg, common borrow, and a 6-inch-thick layer of topsoil. Trees were planted in a portion of the area cleared during the site preparation. All the disturbed areas on-site were subsequently hydroseeded. The original fencing alignment was restored at the conclusion of disposal operations, gated and locked. The background level of lead in soil was found to be 100 mg/kg in areas outside the site. The proper restoration of the site, despite containment of lead-contaminated soils above the background levels (between 1,000 and 100 mg/kg), was effective in the use of BMPs to control migration of contaminants as shown by surface water and groundwater monitoring conducted after site remediation.

11.4.3 Views and discussion

In addition to ambient air quality monitoring throughout the remedial action (see "B." in Section 11.4.2), personal air monitoring was conducted at this site during the startup of each new type of work activity to determine the potential personal exposure to lead. The action level for lead during personal air monitoring was 50 $\mu g/m^3$ for an 8-hour time weighted average (TWA). Both ambient and personal air monitoring were carried out in accordance with the Site Health and Safety Plan (SHSP), to determine that the airborne lead concentrations on- and off-site were within acceptable ranges.

11.5 PARTICLE SIZE SEPARATION AND SOIL WASHING AT A SITE PREVIOUSLY USED FOR AMMUNITION TESTING AND DISPOSAL IN CONNECTICUT (SITE 4)

11.5.1 Description of site remediation activities

Between September and November 1995, a pilot study was conducted at this site to evaluate soil washing and chemical leaching of contaminated soil excavated from several areas of environmental concerns (AECs) in a large industrial property spread over 435 acres of wooded land. A total of 51 AECs, including three RCRA regulated units, were identified in the property. Preliminary evaluation of remedial alternatives had indicated that soil washing may be feasible as a soil treatment method in the majority (37 out of 51) of AECs. It was estimated that approximately 40,000 cubic yards of soil needed treatment at these AECs.

The constituents of concern in the soil were inorganic metals (primarily lead). Organic constituents were not of particular concern because they were generally present at concentrations below the proposed RCRA corrective action levels. Soil composition and contaminant distribution varied significantly among the AECs, and in some cases, within each AEC. A pilot study was therefore required to determine a viable and cost-effective remedy, which would be protective of human health and the environment.

The site applied for a Corrective Action Management Unit (CAMU) status for an area adjacent to one of the AECs to facilitate the storage and processing of soil from various AECs during the pilot study. The EPA approved this request in October 1994 and a paved pad was constructed for management of excavated soil and treated soil at the CAMU. The pilot study was conducted at the CAMU pad and included soil washing runs at varying feed rates (based upon the amount of particulate lead, physical characteristics, and organic material present in the soils) obtained from different AECs. The treatment goal of each soil washing run was to reduce total lead content to 1,000 mg/kg, as well as to meet the TCLP or synthetic precipitation leaching procedure (SPLP) levels for determining the treated soil to be non-hazardous. Chemical leaching was conducted on feed soils with high total lead content (10,000 mg/kg) and on a few batches of residuals from soil washing not meeting the treatment goals for TCLP/SLP. In addition to a pilot study of soil washing/soil leaching, the site conducted a pilot study of processing the shotgun shells separated from the contaminated soil during soil washing pilot runs.

The pilot study found that soil washing provided an optimized and steady-state processing of soil with varying characteristics excavated from the AECs. Based on the results of the pilot study and a comparison of treatment alternatives, soil washing was recommended for full-scale operations at the site. A few process changes and alternative risk-based treatment criteria were also recommended by the study. The existing CAMU has also been recommended, with a few modifications, for use in full-scale soil washing operations at the site.

11.5.2 BMPs used to prevent the cross-media transfer of pollutants

During the pilot study involving soil excavation at AECs and soil treatment operations at the CAMU, the following concerns for cross-media transfer of pollutants existed:

A. Migration of contaminants, such as lead, mercury, strontium, asbestos, etc. as dust from different work areas, and
B. Surface water/groundwater contamination with metals and organics released from the soil during soil washing and chemical leaching operations.

The pilot study was therefore designed to address these concerns while evaluating alternative soil treatment methods for their performance in meeting the treatment goals. Accordingly, BMPs were used during the pilot study. The details on the field application of the following six BMPs, as introduced in different remedial stages, are provided below:

A. Site preparation and staging
   a. Monitoring of air quality at AECs and the CAMU pad:
      During the pilot study, both air sampling and real-time monitoring of air quality were performed for lead, total dust, and other contaminants. Baseline samples

were collected at the perimeter of each area before the start of operations to obtain background data. Thereafter, air sampling was performed over the entire period of operations during each day in the area of interest. The air samples were sent to a laboratory approved by the American Industrial Hygiene Association (AIHA), which used NIOSH or OSHA methods. Typically, air samples were taken downwind of the area of interest. In some cases, samples of air were also taken upwind of the area. Real-time monitoring of total dust in the air was conducted periodically at upwind and downwind locations during soil excavation at AECs. Monitoring of total dust was performed by an aerosol monitor to determine the concentrations of particulates in the air. A photo-ionization detector (PID) was used to screen VOCs in soil periodically during excavation activities at AECs. Real-time monitoring of total dust and mercury was carried out every 30 minutes around the CAMU pad.

b. Installing a new CAMU pad for soil treatment operations:

The CAMU pad was designed to store the following major types of soils/soil fractions (until the completion of sampling and analysis to determine their future management) in full-scale operations: excavated soil, trommel oversize materials, wet-screen oversize materials, "clean" coarse-grained sand, and the concentrated fine-grained materials. The area for storage of concentrated fine-grained material was provided with a concrete base and containment berms. This area was also given adequate slope to a sump capable of collecting excess water or stormwater run-off from the materials. The oversize materials were stored in a lined storage area segregated by an earthen berm. The soil stockpiles were also stored in a lined storage area with an earthen berm. However, this liner was protected from vehicular traffic by a 6-inch layer of sand and a 3-inch layer of gravel. Closure of the staging area for excavated soil would include removal of the gravel, sand, and flexible-membrane liner placed on the ground.

B. Pretreatment

a. Arranging for dust suppression at AECs and the CAMU pad.

The site controlled airborne metals by suppressing dust during soil excavation and during transportation of soil from AECs. This involved spraying water in the excavation areas and on transported soil. Watering trucks were used for dust suppression in remote areas of the site. The trucks used for transporting the soil were also kept covered as a method of dust suppression. The Health & Safety Plan for the pilot study required the provision of adequate dust control measures in the CAMU pad because of the presence of lead in the soil being treated. However, there was no significant emission of dust during the soil washing process itself because the materials were kept wet during the process. The stockpiles of soil in the CAMU pad were either sprayed with water or kept covered with tarps as a dust suppression measure. Water supply to the CAMU pad was provided by a fire hydrant line in the main manufacturing plant.

b. Providing covered and lined roll-offs for managing soil treatment residuals:

The residuals of soil-washing included contaminated fine-grained material, which was stored preferably in covered and lined roll-offs to prevent leaching of metals and to control dust emission.

C. Soil treatment

a. Collecting and treating stormwater run-off from the CAMU pad and wastewater from soil treatment processes:

During soil washing operations in the CAMU pad, the rainwater accumulated in the excavated soil storage area was pumped out into the soil-washing process. The area used for storage of oversize materials generated wastewater during the high pressure water/steam cleaning process. This wastewater was collected in catch basins and treated in a plant existing on-site. Soil particles separated from the oversize materials were processed through the soil-washing system. The stormwater collected in the fine-grained material storage area was analyzed first to determine the level of contamination and then treated either on- or off-site.

D. Post-treatment
   a. Monitoring groundwater – upgradient/downgradient of the site.

   Groundwater sampling was conducted quarterly – before, during, and after completing the pilot study at the site. For this purpose, two existing groundwater monitoring wells downgradient of the site were used. Analytical data compiled for four quarters at these wells showed no evidence of any impact on groundwater quality due to the metals or VOCs released during soil-washing operations conducted for the pilot study.

11.5.3 Views and discussion

Samples from the area and personal samples were also collected in the CAMU pad during the period of soil-washing or shotgun-shell processing operations. The parameters sampled during the pilot study included lead, total dust, mercury, strontium, acetic acid, and asbestos – i.e. all contaminants of concern known to be present in the materials being processed in the CAMU pad.

Based on the initial results of air sampling at the CAMU and AECs, which showed the presence of mercury in the air at elevated levels, real-time monitoring for mercury was conducted at the property fence line for a brief period in the pilot study. A 'gold film' type mercury vapor analyzer was used for this purpose. Readings were taken both upwind and downwind of the potential or suspected sources of mercury emission during the activities within the property. Background mercury data were also collected. The results showed no risk of public exposure due to the release of mercury during site remediation activities.

Based upon the performance of soil treatment methods and field evaluation of cross-media transfer of pollutants during the pilot study, several BMPs were planned to be modified in full-scale operations at this site. For example, it was decided to conduct only real-time monitoring of total dust on a periodic basis, and bi-weekly air sampling for lead and total dust monitoring during full-scale excavation activities at the AECs. Bi-weekly air sampling was also planned to be conducted at the CAMU pad. Only periodic samples from the personal samples were planned to be performed in the CAMU pad for: lead and total dust during soil washing, and for asbestos during shotgun-shell processing operations. It was also decided to eliminate the monitoring for mercury in full-scale operations.

Other BMPs used in the pilot study that may require changes in full-scale operations include the technologies applied for long-term storage of excavated soils and treatment residuals on-site. For example, the use of heavy-duty poly tarpaulin covers secured with sandbags in the pilot study to cover re-wash soils may have to be replaced with other media for covering soils, such as foam coverings, windscreens, and/or water sprays with additives (see Table 13.2 for more details). The full-scale operations at this large site would probably require the monitoring of surface waters in the area, such as creeks, streams, and wetlands.

11.6 SOIL VAPOR EXTRACTION OF SOLVENTS AT A CLOSED ELECTRONIC COMPONENT MANUFACTURING FACILITY IN MAINE (SITE 5)

11.6.1 Description of site remediation activities

From March 1995 onwards, except for a shutdown during the coldest periods of winter in 1995 and 1996, a full-scale system for soil-vapor extraction (SVE) has been in operation at this site to treat the soil below the main building at the facility. The soil contamination occurred during previous manufacturing operations when spent solvents leaked from a corroded piping used for transporting the solvent wastes to an underground tank outside the building. The soil below the main building was found to contain several VOCs like trichloroethylene (TCE), tetrachloroethene (PCE), ethylbenzene, xylenes, and 1,1,1-trichloroethane. During the last several months of operation, the concentration of VOCs in the influent to the SVE system has reduced from 1,200 ppm to less than 25 ppm. The off-gases generated by the SVE system are processed through granular activated carbon (GAC), which is then regenerated on-site in a mobile steam unit. The steam stripping also recovered light and heavy DNAPLs originally contained in the off-gases. The full-scale system was installed after conducting a range of on-site investigations (e.g., soil borings and soil gas survey, surface water and groundwater sampling, and records search of chemicals used and building constructions at the site). An SVE pilot program was also conducted on-site before starting the full-scale operations.

11.6.2 BMPs used to prevent the cross-media transfer of pollutants

There was only a limited opportunity for cross-media transfer of pollutants from the operation of SVE at this site due to the confinement of contaminated soil and off-gases below the main slab of the existing building. However, SVE pilot testing at this site helped to better define the release potential for off-gases at various points around the facility and the need for proper management of the contaminants (VOCs and DNAPLs) in the off-gases generated by the SVE system. Accordingly, three BMPs were used during this field application. The details of these BMPs used in different remedial stages are provided below:

A. Site preparation and staging
   a. Containerization of drill cuttings:
      The soil monitoring points and SVE wells installed on-site involved the excavation of contaminated soils. To control the emission of VOCs during these activities, soil borings and drilling were performed with appropriate containerization of cuttings.
   b. Ambient and workspace air monitoring for VOCs:
      Ambient air monitoring for VOCs was performed with a photoionization detector (PID) to ensure that the state of Maine's interim guidelines are being met for the maximum ambient air concentrations of TCE (2 $\mu g/m^3$), PCE (0.1 $\mu g/m^3$), ethylbenzene (1,000 $\mu g/m^3$), xylenes (300 $\mu g/m^3$), and 1,1,1-trichloroethane (1,000 $\mu g/m^3$). This activity also enabled a check of emission control measures installed in the SVE system. Workspace air monitoring of VOCs was carried out with a PID to ensure

the protection of worker health and safety. By comparing the workspace air quality with the influent concentration of VOCs, it was possible to check the general performance of the SVE system.

B. Soil treatment
   a. Recovering DNAPLs from soil-vapor:
      By recovering DNAPLs from spent GAC, the site prevented cross-media transfer of these pollutants from treatment residuals.

11.6.3 Views and discussion

Because of the confinement of contaminated soil, and the creation of a negative pressure under the main slab of the building due to SVE application, there was very little potential for the release of VOCs inside the building. Workspace air monitoring was, however, conducted to confirm that the SVE system was operating properly at all times. The confinement of soil under the building also prevented the leaching of contaminants to groundwater or surface water, thus eliminating the need to monitor these media during the operation of SVE at this site.

11.7 ON-SITE CONTAINMENT OF SOILS IN FORMER MANUFACTURING AREAS AT A CHROMIUM PLANT, MARYLAND (SITE 6)

11.7.1 Description of site remediation activities

In 1991, a Corrective Measures Implementation Program Plan (CMIPP) was submitted by the owner of a 140-year-old chromium ore processing facility in Maryland to prevent further migration of contaminants to the soil, groundwater, and surface water. The owner of the facility had stopped manufacturing operations in 1985 and entered into a Consent Decree with the EPA and the Maryland Department of the Environment (MDE) in 1989 to investigate the nature and extent of contamination at the 20-acre facility and submit the CMIPP based on the findings of their investigations. These investigations found that the soils were contaminated with hexavalent chromium (above the action level of 10 parts per million) in a few areas of the site where the maximum concentration of chromium was found to be 94 mg/kg (ppm). Higher levels of contamination were also found in the sediments in the harbor surrounding the peninsular facility. Both the shallow aquifer (0–20 feet below the ground surface) and deep aquifer (23–70 feet below the ground surface) were found to be contaminated with chromium. The highest concentrations were found to be 14,500 mg/L for the shallow aquifer and 8,000 mg/L for the deep aquifer, near the former manufacturing area at the facility. Chromium in the deep aquifer had migrated approximately 2,750 feet off-site along the top of the bedrock where the concentration of chromium was 1,600 mg/L. However, no user of the deep aquifer for drinking water could be identified. The surface water in the marine harbor surrounding the facility was found to be contaminated with chromium at the maximum concentration of 3,170 :g/L (ppb), which exceeded the acceptable level of 50 ppb for chromium in marine waters.

The proposed corrective measures include:

- installation of a deep vertical hydraulic barrier (slurry wall) as a containment structure to prevent the migration of contamination into the marine harbor and groundwater surrounding the facility;
- installation and operation of a groundwater withdrawal system within the containment structure to maintain an inward hydraulic gradient of groundwater at the site;
- construction of a multi-media cap over the containment area to prevent any future exposure to the contaminated soil and control the generation of contaminated leachate from any infiltration of precipitation at the site; and
- monitoring the surface and groundwater in a comprehensive manner to confirm that all the site remediation goals were being achieved.

In preparation for implementing these corrective measures, the owner dismantled the manufacturing plant existing at the facility, which included - 21 buildings, 240,000 square feet of transite roofing panels, 15,000 tons of decontaminated equipment and structural debris, and 50,000 tons of concrete and rubble. All of these were sampled, classified, and shipped to appropriate off-site facilities. Before the construction of containment structures, approximately 150,000 cubic yards of sediments were dredged from the surrounding facility at the harbor, a new outward embankment was constructed, and soil found on-site contaminated with chromium or other hazardous contaminants above the action levels was excavated and disposed of off-site.

The implementation of the CMIPP commenced in 1993 after completely dismantling the plant and providing asphalt cover over the former manufacturing areas at the site. The deep vertical hydraulic barrier was later provided with a slurry wall mixture of soil and bentonite encompassing 15 acres of the site. This slurry wall was trenched over a linear distance of 3,300 feet by 3 feet wide with a depth ranging from 65 to 80 feet to the bedrock. A multi-media cap of geosynthetic clay liner and 60 mil LDPE geomembrane covers were provided over the area within the slurry wall. A system for pumping, treating, and disposing of the groundwater from the site is operational. This system is designed to operate in order to maintain an inward hydraulic gradient of 0.01 foot per foot from outside to inside the slurry wall. After completing the construction of a multi-media cap, the concentration of chromium in the surface water outside the facility will be maintained within regulatory levels by monitoring the performance of the slurry wall and by controlling the rate of groundwater extraction from the site. This site remediation was designed to permit future development of the site as a mixed-use (recreational and commercial uses) zone.

11.7.2 BMPs used to prevent the cross-media transfer of pollutants

The RCRA corrective action nearing completion at this site is designed primarily to minimize the future releases of contaminants from the soils to the air, surface water, and groundwater. In addition, the site remediation activities were conducted in a manner to prevent any significant cross-media transfer of pollutants during site preparation and installation of containment structures at the site. The containment structures were also designed to minimize the possibility of improper operation during the future development and use of the site for recreational or commercial purposes. Accordingly, 11 BMPs were used during this field application. The details of these BMPs used in different remedial stages are provided below:

A. Site preparation and staging
   a. Perimeter and personnel monitoring of air quality:

      Before implementing the CMIPP, the existing buildings, equipment, and structures at the facility were dismantled under a plan approved by the MDE and incorporated into the Consent Decree. This plan required the perimeter monitoring of air quality for the possible release of chromium and asbestos during dismantling operations. The facility installed six air-sampling stations on the perimeter of the site, which operated continuously for 24 hours to filter the ambient air and provide samples for analysis of the concentrations of chromium and asbestos. The air quality was then compared with the standards for average and maximum concentrations of each pollutant. The concentration of a pollutant as measured by any of the six air-sampling stations was also normally expected to be within twice the standard deviation of the recent values as measured by all six stations. Otherwise, it was assumed that an exceedance of air quality standard had occurred and that a corrective action was required. In addition to monitoring the air quality on the perimeter of the site, personnel at the site were also monitored for their exposure to chromium and asbestos to assure the health and safety of personnel working on-site during the dismantling and disposal of the plant.

B. Pre-treatment
   a. Monitoring trends in air quality and weather conditions to control construction activities on-site:

      The facility monitored any trends in the concentration of chromium in ambient air on a daily basis. These trends were examined rather than waiting for an exceedance of air quality standards taking place on-site. These trends were discussed at daily meetings and a possible list of reasons prepared for any trends observed. These were also compared with the related set of weather conditions as measured by a weather station installed on-site (e.g., wind speed and direction were compared with trends in the concentration of chromium). In a specific case of air quality monitoring at this site, it was found that high concentrations of chromium were probably related to a spell of dry and windy weather. In response, new efforts were made to suppress dust emission by spraying water on the stockpiles and other areas of construction on-site. Work has also been stopped on several occasions when air quality standards were exceeded.

   b. Covering debris generated during construction:

      A site visit during the regrading of the site before the placement of a multi-media cap showed that the piles of debris generated during previous construction activities were kept covered under sheets of plastic. This practice was followed on the site mainly to guard against the migration of any dust or other debris from open piles to the harbor nearby, which is commonly used for recreation.

   c. Providing temporary sumps for collecting stormwater run-off from the site during construction:

      During the same visit, it was found that temporary sumps were provided with a pumping system to collect and transfer any run-off from the site to the tanks being used for storing groundwater extracted on-site. This arrangement prevented a transfer of site pollutants to the surface water during construction. After construction of the cap, a permanent system was planned to be made available for diverting stormwater run-on to the site and collecting stormwater run-off from the site.

d. Preventing surface water pollution during construction of the slurry wall:

In addition to the detailed specifications and inspections required to assure a high-quality construction of the slurry wall and trench at this site, a few precautions were taken to prevent the cross-media transfer of pollutants during construction. For example, the construction spoils from the tanks were placed above the 100-year level of high tide, and were also covered by a sheet of plastic. These spoils were tested for the presence of high concentrations of chromium, and appropriate management of stormwater run-on/run-off was provided. Fugitive dust emission from the spoils was controlled during periods of dry weather by sprinkling water over the spoils.

e. Providing temporary culverts for trucks crossing over the slurry walls during the remaining construction:

During the construction of the multi-media cap over the site, a short bridge of concrete was provided over the slurry wall at several points to permit the occasional movement of trucks. This arrangement prevented damage to the slurry wall so that its containment performance was unaffected.

C. Soil treatment

a. Checking the integrity of slurry wall:

The containment performance of the slurry wall was assessed after its construction (but before the final remedial construction on-site) using a series of hydraulic tests and monitoring. For testing purposes, paired piezometers were designed to the same specification as those of the final groundwater extraction wells. Water levels in the shallow aquifer outside the slurry wall rose at an average rate of 0.35 feet of head per month during and immediately after slurry wall construction. Individual pumping tests were performed in the deep aquifer and at four locations inside the site perimeter. In these locations, even with 25 feet of drawdown, the outside piezometers did not indicate the influence of the pumping well. Thus, barrier integrity was confirmed in the vicinity of the pumping and monitoring. Several interior piezometers were then pumped simultaneously to simulate the groundwater withdrawal after remedial construction on-site. Tests confirming earlier pumping test results showed rapid drawdown propagation in the confined aquifer within the slurry wall. Steel plates were embedded in the slurry wall at several locations and used as a level gauge to visually monitor any settlement of slurry wall contents, and as a direct measurement of subsidence after construction.

b. Providing standby well-heads for additional future groundwater extraction on-site:

In anticipation of future problems in operating the groundwater pumping system at the specified locations, the site had provided standby well-heads (without pumps) which could be used as a contingency. This feature minimizes the need to damage the multi-media cap and drill new wells for an upgrade of the groundwater extraction system after site development.

c. Providing a capillary break layer in the multi-media cap:

The site used a capillary break layer to prevent any capillary rise of contaminated water from the site to the low-permeability layer (containing geosynthetic clay liner and geomembrane) above. Upward migration of contaminants was thus prevented.

d. Preparing multi-media cap for future site development:

As the site was permitted for development as a multi-use zone, concept designs were prepared for the cap in areas to be paved or unpaved in the future. A brightly colored (orange) geotextile was placed at 18 inches beneath the surface of the multi-media cap with grass surface cover. This was placed to alert future developers of the site that a penetration of the cap below this point might result in infiltration of water or any other liquids to contaminated soils below the cap.

D. Post-treatment

a. Environmental monitoring plan for checking the effectiveness of site containment:

Monitoring of chromium in surface water and groundwater levels both inside and outside the slurry wall was planned to be continued regularly after completing the installation of multi-media cap and groundwater drainage wells. It was expected that the standards of 50 ppb for chromium in surface water and the maintenance of the required gradient in groundwater will be achieved in the future by pumping and treating groundwater at the rate of only about 2,000 gallons per day (gpd). In contrast with this rate of groundwater withdrawal, the temporary arrangements on-site during the construction were pumped at about 60,000 gpd.

11.7.3 Views and discussion

While preparing this site for containment, it was necessary to take up other large tasks which were, by themselves, major remedial actions and included the use of additional BMPs to prevent the cross-media transfer of pollutants. These tasks include: (1) dredging of contaminated sediments from the harbor and replacing the dredged sediments with clean stone for stabilizing the existing bulkheads around the site and (2) dismantling the old chromium plant under negative pressure to prevent the emission of chromium and asbestos into the neighboring areas as fugitive dust.

Because of the high levels of contaminant encountered, the dredging and disposal of the sediments were accomplished under stringent environmental controls. To ensure that the sediments with excessive concentrations of heavy metal were not taken to the disposal facility off-site, testing and bulking of every load of dredged spoils was required. Dredging was also performed completely within a turbidity curtain. Water sampling and analysis for chromium were conducted inside and outside the curtain to check the curtain's effectiveness in reducing migration of chromium in the surrounding harbor waters. In one area of the site where confined space would have made dredging very problematic, the sediments were stabilized and capped in place after the construction of the rock embankment at this location.

Before constructing the slurry wall, a new rock embankment was constructed to prevent any unexpected collapse of contaminated soils into the harbor during the construction of a trench for the slurry wall. This embankment was located outside the existing bulkheads along the boundary of the site with the marine harbor. The new outboard embankment also enabled the containment of the contaminated surfaces of the bulkheads within the slurry wall.

The dismantling of the plant was conducted with a series of controls designed to assure worker health and safety during these operations. The buildings at the site were categorized according to concerned pollutants (i.e. only asbestos, asbestos and chromium, and only chromium). The dismantling plan also required the development and fabrication of enclosures to create a negative pressure during the dismantlement of some buildings. One of the buildings, for example, used nine HEPA filters with a capacity of 18,000 cfm each, as well as the use of water curtains and air seals during dismantling operations. This plant building was 300 feet long and 70 feet wide, with the maximum height of the roof being 100 feet. There was another large building with similar dimensions and several smaller buildings that were dismantled under negative air pressure.

11.8 EX SITU BIOREMEDIATION OF EXPLOSIVES CONTAMINATED SOILS AT A DOD FACILITY IN VIRGINIA (SITE 7)

11.8.1 Description of site remediation activities

In 1996, a Superfund removal action involving the excavation and transportation of 800 cubic yards of explosives-contaminated soil was completed within three days at one of several sites within a DoD facility being investigated under an Installation Restoration Program (IRP). At this site, the contaminated soils were found in a drainage area located near wetlands and along a small tributary within the Chesapeake Bay watershed. The site had received nitramine-containing wastewater from a weapons manufacturing plant since 1945. Although this effluent was diverted in 1986 to a sanitary sewer and the site had reverted to a natural drainage area, explosive compounds such as TNT, HMX, and RDX were found at elevated concentrations in the samples of soil and sediments obtained recently from the area. A decision was made to excavate soils up to 4 feet depth from the drainage area, which was partly subject to tidal action every day. Parts of the drainage area were on a wooded slope of about 15 degrees leading to the wetlands.

After clearing the wood and preparing the site for excavation, two excavators were operated round-the-clock to remove and load contaminated soil to dump trucks parked on the road, about 20 feet above the bottom of the drainage area. The excavated soil was transported over a distance of less than 1 mile to another location where a suitable biocell had been constructed earlier to conduct a pilot study of an anaerobic process for treating explosives in the soil. Approximately 600 cubic yards of the explosives-contaminated soil were screened and then slurried before being pumped into the biocell for treatment. Proper operating conditions were maintained in this biocell and the treatment goals for removal of explosive compounds from the soil and supernatant were achieved, as planned, within nearly 60 days of bioremediation. The treated biocell contents were left in place and were planned to be covered with topsoil and vegetation after allowing excess water to evaporate. Approximately 200 cubic yards of untreated soil were removed from the contaminated site, and over 1,000 cubic yards of soil excavated to construct the biocell were remaining for disposal in the pilot study area.

As the results of the pilot study were positive and treatment goals seem to have been achieved, scaled-up operations using the anaerobic bioremediation process were considered for treating explosives contaminated soils found at other sites within the facility.

11.8.2 BMPs used to prevent the cross-media transfer of pollutants

The pilot study of soil bioremediation and removal action completed at this site used BMPs to prevent a few likely incidents of cross-media transfer of pollutants, such as surface water pollution due to stormwater run-off/run-on. However, there were other concerns, which would have to be addressed by BMPs in full-scale operations at this site. The details of five BMPs used at this site during the pre-treatment activities to prevent the cross-media transfer of contaminants are provided below:

A. Pretreatment
   a. Arranging for containment of soils during excavation:
      During site preparation, a silt fence and straw bales were installed downslope of the excavated area along the border of the site with the wetlands nearby. These soil erosion and sedimentation controls prevented, to some extent, the carry-over of contaminated soils and sediments by daily tidal action on the site. The rapid completion of soil removal within three days also helped with better containment of pollutants at the site.
   b. Controlling the transportation of excavated soils:
      The dump trucks used for transporting excavated soil from the site being remediated to the biocell were loaded by excavators outside the site. A fixed route was always used for transportation to limit the areas outside the site from being contaminated by spills on the road. Roll-off containers were used instead of dump trucks to transport wet soils excavated from the site.
   c. Providing for stormwater run-on/run-off control near biocell:
      The biocell area was on a plateau and existing run-on was diverted around the area via a drainage swale and an existing culvert. In addition, the biocell was given an adequate (18–24 inches) freeboard to accept direct precipitation and any stormwater run-on during normal operations. At the same time, two portable tanks were kept ready on-site to pump out additional water draining into the biocell during storms.
   d. Using soil erosion and sedimentation controls near biocell:
      Stormwater runoff from the biocell was designed to pass through soil erosion and sedimentation controls. A silt fence and straw bale check dams were therefore installed on the downslope of the area used for the biocell. These controls were placed before starting the construction of the biocell. The area was then graded to ensure that all stormwater runoff passes through the controls, which were planned to remain until the site is vegetated after completing all operations. Any vehicular traffic in the site was limited to the haul roads leading to the biocell.
   e. Containing spills during pre-treatment operations:
      Any leaks or spills of contaminated water during the screening and slurrying of contaminated soils were automatically drained by the gradient into the biocell. This prevented the contamination of soils outside the biocell.

11.8.3 Views and discussion

The full-scale site remediation based on this pilot study should consider the use of additional BMPs during site preparation and staging, pre-treatment and soil treatment. The site had already planned a facility-wide groundwater monitoring and treatment in lieu of the post-treatment phase of all site cleanup efforts. Additional BMPs that could be used in full-scale operations include the following:

- An assessment of the potential impact on the ecology of neighboring areas should be performed before conducting excavation for ex situ remediation of other contaminated sites that are located near the wetlands. The feasibility of in situ bioremediation should be considered as an alternative to prevent the cross-media transfer of pollutants during excavation and transportation, as well as the potential for contaminating the new site selected for operating the biocell.

- The site should be well protected from flooding during storms.
- A storage area (with liners, berms, and top cover) should be provided for excavated soils to prevent leaching of contaminants from the soils awaiting treatment.
- Future efforts using ex situ bioremediation should provide excess capacity for treatment to prevent stockpiling of untreated soil on-site at the end of operations. Marshy conditions usually make it necessary to excavate more quantities of soil than planned and the excavated soil may contain large quantities of fine tree roots, which cannot be removed easily from the soil before treatment.
- Surface water run-off should be monitored during excavation on marshy land subject to daily tidal action.
- The air emission of organics (e.g., methane) should be monitored during treatment in the biocell. A log should be maintained to record air monitoring data with details of wind direction on a daily basis during full-scale operations.
- The use of a dragline instead of excavators should be considered as alternative equipment in marshy lands to avoid the problems of getting stuck in marshy areas (as experienced at this site).
- Biocells used in large operations should be equipped with concrete floors with protective lining systems, such as studded liners, instead of the flexible double liners separated by sand as used in the pilot.
- The pipes used for recirculating process water in the biocell should be run within the unit to prevent contamination of outside soil from leaks in these pipes.

The following three case studies focus on:

- Remedial efficacy,
- Regulatory requirements/cleanup goal,
- Innovative and combination of technologies (technology train), and
- Costs

11.9 GROUNDWATER SPARGING AT AMCOR PRECAST IN OGDEN, UTAH (SITE 8)

11.9.1 Description of site remediation activities

Amcor Precast in Ogden, Utah, stored gasoline and diesel fuel in three underground storage tanks (USEPA 2000). A release was discovered in 1990. An investigation in 1991 indicated that the areal extent of groundwater contamination was approximately 2,789 m^2 (30,000 ft^2) and that an estimated 5,126–5,356 m^3 (6,700–7,000 yd^3) of soil had been contaminated. The primary contaminants of concern were benzene, toluene, ethylbenzene, and xylenes (BTEX), naphthalene, and total petroleum hydrocarbons (TPH). A density-driven groundwater sparging system and soil vapor extraction (SVE) system were installed in January/February 1992 and operated from March 1992 to September 1993. The sparging system was used as the primary remediation technology. SVE was used locally to treat volatilized hydrocarbons, created by the air stripping process, and prevent contaminants from migrating to nearby office buildings.

With the density-driven groundwater sparging system at Amcor, water inside the wellbore was aerated by injecting air into the base of the wellbore (rather than injected under pressure) with the resulting injection of air bubbles stripping contaminants from the water while increasing the dissolved oxygen content. In addition, the aeration process acted to create groundwater circulation and transport. Therefore, with this system, petroleum hydrocarbons were removed from the subsurface by -

1 . Aerobic biodegradation resulting from the supply of oxygen to the saturated zone and
2. In situ air stripping.

The vapors from air stripping are transferred to the vadose zone and are biodegraded in place. The application of density-driven groundwater sparging and SVE achieved the specified cleanup goals for both soil and groundwater. The cleanup goals for soil and for all contaminants except naphthalene in the groundwater were achieved within 11 months and for naphthalene in groundwater within 18 months of system operation.

The total capital cost for this application was about $157,000 (current value in 2020- US $267,282) and total annual operating costs were $62,750 (2020 value - US $106,823). Air sparging is limited to contaminants that can be degraded by indigenous bacteria under aerobic conditions. Maximum sparging well airflow and groundwater wellbore circulation rates are dependent on well diameter, depth to groundwater, and the hydraulic conductivity of the formation. Therefore, longer remediation times or a greater number of sparging wells may be required in lower permeability formations (Dutta, 2002).

11.9.2 Remedial efficacy, cleanup goal, and other factors

The remedial efficacies and other factors of this case study are summarized as follows:

In situ density-driven groundwater sparging and SVE were used at this site. The cleanup system used at this site consisted of three main components:

- Groundwater sparging system;
- Groundwater recirculation system; and
- SVE system

Groundwater sparging (SVE was used locally)	*Principal method of remediation;* Sparging System – Density-driven groundwater sparging – removed petroleum hydrocarbons using (1)aerobic degradation and (2) in situ air stripping; water inside the wellbore was aerated directly by injecting air at the base of the wellbore – 12 groundwater sparging wells installed to a depth of 18 feet.
Groundwater recirculation	Three downgradient extraction (pumping) wells were installed to a depth of 20 feet and one upgradient injection galley (former tank excavation backfilled with pea gravel).

Soil vapor extraction	Three vertical extraction wells located adjacent to the pumping wells – Vapor discharged to atmosphere.
Cleanup authority: (state)	Utah Department of Environmental Quality, Division of Response and Remediation (DERR)
Contaminants:	Benzene, Toluene, Ethylbenzene, and Xylenes (BTEX), Naphthalene, and Total Petroleum Hydrocarbons (TPH)
Concentrations:	*Groundwater*: Average groundwater concentrations (mg/L) in plume area/site maximum – TPH (51/190), benzene (1.3/4.7), toluene (2.4/9.4), ethylbenzene (0.78/2.7), total xylenes (2.5/8.0), naphthalene (0.18/0.63) *Soil*: Average soil concentrations (mg/kg) in plume area/site maximum – TPH (555/1,600), benzene (2.0/7.8), toluene (1.4/2.5), ethylbenzene (5.7/19), total xylenes (37/110)
Waste source:	Underground Storage Tanks
Type/quantity of media treated:	Groundwater and Soil. The site stratigraphy consisted of interbedded silty sand and poorly graded fine gravel underlain by a silty clay aquitard at a depth of approximately 18 feet below ground surface. Depth to groundwater ranged from 5 to 11 feet; aquifer thickness (7–13 feet) – Porosity (20–35%), hydraulic conductivity (190 ft/day) – Aerial extent of the plume – approximately 30,000 ft^2; and the vertical extent of contamination - ranged from approximately 5–11 feet below ground surface. Thus, the contaminated volume is estimated at 5,350 m^3 or 7,000 yd^3.
Purpose/significance of application:	Full-scale remediation of groundwater contaminated with diesel and gasoline fuels using in situ density-driven groundwater sparging and soil vapor extraction.
Regulatory requirements/cleanup goals:	*Soil*: Utah Department of Environmental Quality (DEQ) Recommended Cleanup Levels (RCLs) – TPH - 30 mg/kg; Benzene - 0.2 mg/kg; Toluene - 100 mg/kg; Ethylbenzene - 70 mg/kg; Xylenes - 1,000 mg/kg; Naphthalene - 2.0 mg/kg. *Groundwater*: BTEX and naphthalene to below MCLs; no cleanup goal for TPH in groundwater *Air*: no air discharge permit was required because air emissions were below de minimis standards of the Utah Division of Air Quality.
Results:	The cleanup goals were achieved for all contaminants of concern in both soil and groundwater.
Cost factors:	*Total capital cost*: $157,000 (or **$267,282–**2020 value), including drill/install wells and sparging system, start-up, project management – *Total annual operating cost*: $62,750 **($106,823–**2020 value) (including electricity, maintenance, and monitoring)

11.10 THERMAL DESORPTION AT THE ARLINGTON BLENDING AND PACKAGING SUPERFUND SITE, ARLINGTON, TENNESSEE (SITE 9)

11.10.1 Description of site remediation activities

The Arlington Blending and Packaging Superfund site, located in Arlington, Tennessee, is a 2.3 acre site that was used for the formulation and packaging of pesticides and herbicides from

1971 to 1978. Chemicals handled at the facility included the pesticides endrin, aldrin, dieldrin, chlordane, heptachlor, lindane, methyl parathion, and thimet as well as solvents and emulsifiers used in the preparation of formulations of these chemicals. Leaks and spills of chemicals occurred during these operations and process water from the facility was discharged to drainage ditches at the site. The site was placed on the National Priorities List (NPL) in July 1987. A remedial investigation (RI), begun in 1988, determined that the main areas of soil contamination at the site were located around and beneath the process buildings. The record of decision (ROD), signed in 1991, specified excavation of contaminated soil and treatment on-site using thermal desorption. Smith's low temperature thermal aeration (LTTA) process was used to treat the contaminated soil at the site. The unit included a direct-fired rotating dryer that heated the soil using a hot air stream. The heated soil was discharged from the rotary dryer to an enclosed pug mill where it was cooled and re-humidified by quenching with water. The treated soil was then sampled, and based on the results, backfilled on-site or stabilized and shipped off-site for disposal. A total of 41,431 tons of contaminated soil in 84 batches were treated during this application. All but six batches of soil met the cleanup goals for the organics on the first pass through the system. Three batches exceeded the cleanup levels and were retreated. Three additional batches slightly exceeded the cleanup goal for total chlordane. Based on the concentrations, the EPA determined that the batches did not have to be treated any further. Following confirmation that the cleanup goals had been met, treated soil was backfilled at the site. Only one batch of treated soil did not meet the total arsenic limit and was shipped offsite for disposal in a Subtitle C landfill.

The original estimate for the soil excavation was 10,000 tons, based on the results from field-based screening using the Drexel method. Subsequent verification analyses indicated that the results from this method were not accurate. The site was recharacterized, using immunoassay sampling (results confirmed to be accurate by an off-site laboratory), and an additional 30,000 tons of soil requiring excavation were identified. The use of immunoassay sampling saved time by providing real-time results.

11.10.2 Remedial efficacy, cleanup goal, and other factors

The remedial efficacies and other factors of this case study are summarized as follows:

The low-temperature thermal desorption technology was used to remediate the contaminated soils at this site. By using a hot air stream from a direct-fired rotating dryer the contaminated soils were heated to a temperature of 343°C–449°C (580°F–750°F). The following main components were used:

- Propane gas was used to heat the air stream, and the organic constituents in the soil were desorbed in the dryer through contact with the heated air.
- Off-gas treatment included a cyclone/baghouse system; a low pressure drop Venturi air scrubber; and vapor-phase carbon adsorption.
- A vacuum of 0.24–0.46 cm (0.10–0.18 inch) of water was maintained throughout the process train.

Low-temperature thermal desorption **Cleanup authority: Federal (EPA)**	Principal method of remediation: Direct-fired rotating dryer heated the soil using a hot air stream Cleanup Authority: CERCLA – Remedial Action – Record of Decision (ROD) signed June 28, 1991

Contaminants:	Pesticides and Metals
Concentrations:	Maximum concentrations during remedial investigation: chlordane (390 mg/kg surface and 120 mg/kg subsurface); endrin (70 mg/kg surface and 20 mg/kg subsurface); pentachlorophenol (130 mg/kg surface and 9.5 mg/kg subsurface); arsenic (370 mg/kg surface).
Waste source:	Leaks and spills of pesticides during blending and packaging operations; process wastewater discharged to drainage ditches at the site.
Type/quantity:	Soil -41,431 tons
Media treated:	Soils primarily silty sands with an average moisture content of 17 wt%; pH of soil -6.8 SU (Standard Unit).
Purpose/significance of application:	Application of low-temperature thermal desorption to treat pesticide-contaminated soil.
Regulatory requirements/cleanup goals:	*Cleanup goals for organics were*: chlordane (3.3 mg/kg); heptachlor (0.3 mg/kg); pentachlorophenol (0.635 mg/kg); endrin (0.608 mg/kg); heptachlor epoxide (0.2 mg/kg) *Cleanup goal for arsenic* initially established at 25 mg/kg in ROD; changed to 100 mg/kg. in explanation of Significant Differences (ESD) All treated soil with a total arsenic concentration >100 mg/kg was to be disposed of off-site. Any treated soil with total arsenic concentrations >100 mg/kg and leachable arsenic >5 mg/L (determined by the toxicity characteristic leaching procedure) was required to be identified as hazardous waste and stabilized before disposal off-site. *Emission standards for the unit*: total hydrocarbons (500 ppmv); particulates (0.08 gr/dscf); and system removal efficiency (>95%).
Results:	A total of 84 batches of soil (41,431 tons) were treated. All but six batches of soil met the cleanup goals for the organics on the first pass through the system. Three batches exceeded the cleanup levels and were re-treated and met the cleanup goals. An additional three batches were slightly above the cleanup levels for total chlordane. Based on the concentrations, the EPA determined that the batches were not required to be re-treated. One batch of treated soil did not meet the 100 mg/kg limit for arsenic and was shipped offsite for disposal in a Subtitle C landfill; however, because the TCLP level for arsenic was below the 5 mg/L limit, solidification/stabilization before off-site disposal was not required. Compliance with the emissions standards was verified during the performance test. The unit met all emissions standards during the three test runs, achieving a system removal efficiency >99.999%.
Cost factors:	Total project cost was $5,586,376 (**$9,510,431–**2020 value), including $4,356,244 (**$7,416,214–**2020 value) in costs directly associated with the thermal treatment. Treatment costs included $4,293,893(**$7,310,065–**2020 value) in capital costs and $62,351(**106,148–**2020 value) in O&M costs. The calculated unit cost for this application was $105 (**$179–**2020 value) per ton, based on 41,431 tons of soil treated.

11.11 IN SITU BIOREMEDIATION USING MOLASSES INJECTION AT THE AVCO LYCOMING SUPERFUND SITE, WILLIAMSPORT, PENNSYLVANIA (SITE 10)

11.11.1 Description of site remediation activities

The Avco Lycoming Superfund site (Lycoming) is a 28-acre facility located in Williamsport, Pennsylvania, US. Since 1929, various manufacturing companies have operated at the site. Past waste handling practices have contaminated the site, including disposal of waste in wells and lagoons, and spillage and dumping of wastes from metal plating operations.

In 1984, the state identified volatile organic compound (VOC) contamination in the local municipal water authority well field located 3,000 feet south of the site. A pump and treat system was installed in the mid-1980s. In May 1995, the PRP proposed the use of in situ bioremediation to replace the pump and treat remedy. Pilot studies of molasses injection and air sparging/SVE were conducted from October 1995 to June 1996. A new ROD, issued in December 1996, replaced the pump and treat remedy with in situ bioremediation, and a full-scale system has been operating at the site since January 1997. Construction of the air sparging/SVE system was suspended in the spring of 1998, due to higher than anticipated water levels.

The use of molasses injection was shown to create an anaerobic reactive zone in an 18-month period where concentrations of TCE, DCE, and hexavalent chromium were reduced. According to the contractor of the Principal Responsible Party (PRP), this technology was shown to save substantial resources when compared to pump and treat.

11.11.2 Remedial efficacy, cleanup goal, and other factors

The remedial efficacies and other factors of this pilot project followed by the full-scale cleanup are summarized as follows:

The In situ Bioremediation clutched with an Anaerobic Reductive Dechlorination treatment system was first started as a pilot study at this Superfund site from October 1995 to March 1996. After successful demonstration of the pilot system, the full-scale system was installed to remediate contaminated groundwater. Ongoing performance data were made available through July 1998.

The major components and factors are listed below:

In situ bioremediation; with an anaerobic reductive dechlorination system	Principal method of remediation: • Pilot studies consisted of molasses injection and air sparging/soil vapor extraction • Full-scale molasses injection system consisted of 20 injection wells, four-inches diameter, ranging in depth from 19 to 30 feet, completed in the overburden • Molasses was added two times a day at variable concentrations and rates • Eight additional wells were used for monitoring system performance • This proprietary technology was owned by ARCADIS Geraghty & Miller

Cleanup authority:	CERCLA – ROD signed December 1996

Contaminants: Concentrations:	• Chlorinated solvents and heavy metals – TCE, DCE, VC, hexavalent chromium, cadmium • Maximum concentrations measured in late 1996 were TCE -700 μg/L, hexavalent chromium -3,000 μg/L, and cadmium -800 μg/L.
Waste source:	Spills and leaks from plating operations; disposal in lagoons and wells

Type/quantity media treated:	Groundwater – estimated at 1.5 Million Gallons; and Soils • Site geology consists of a sandy silt overburden overlying a fractured bedrock and a fractured limestone • Target area for treatment is the shallow overburden to approximately 25 feet below ground surface, covering approximately 2 acres.
Purpose/significance of application:	One of the first applications of molasses injection technology on a full scale at a Superfund site.
Regulatory requirements/ cleanup goals:	The 1996 ROD specified the following cleanup goals for groundwater: TCE - 5 Φg/L; 1,2-DCE - 70 Φg/L; VC -2 Φg/L; Cd - 3 Φg/L; Cr^{+6} - 32 Φg/L; Mn - 50 Φg/L.

Results:	• The pilot study showed that the technology was able to create strongly reducing conditions. • The baseline sampling event showed that anaerobic, reducing conditions were present only near two of the site monitoring wells. • Since the injection of reagent, the redox levels have decreased to anaerobic conditions in many of the wells that had previously indicated an aerobic environment, and cleanup goals have been met in some of the wells. • Analytical results for TCE, DCE, and VC for an area that was converted from aerobic to anaerobic show that TCE was reduced from 67 to 6.7 :g/L, a 90% reduction. The concentration of DCE initially increased, indicating the successful dechlorination of TCE, and then decreased to 19 :g/L. • Concentrations of TCE, DCE, and Cr^{+6} have been reduced to less than their cleanup goals in many of the monitoring wells at the site.
Cost factors:	ARCADIS Geraghty & Miller reported a total project value of $145,000 ($246,852–2020 value) for the pilot study application at this site, including the preparation of a work plan. The costs for the construction of the full-scale molasses injection system was approximately $220,000 (374,535–2020 value). Operation and maintenance, including monitoring was approximately $50,000 **($85,122–**2020 value) per year.

11.12 COMPARISON OF CASE STUDIES ON BMPs USED VERSUS THE RECOMMENDED BMPs (USEPA, 1997)

According to the field validation study by EPA for comparison of the BMPs recommended in the guidance (USEPA, 1997) versus the BMPs used in the field, it was found that some BMPs were used at the selected sites to control cross-media transfers in different remedial phases. In a few selected case studies, EPA found that additional BMPs could have been used, which have been identified in each specific case study under views and discussion earlier in this chapter.

The case studies covered in this chapter also indicated that most BMPs are introduced in the earlier two phases of site remediation, i.e. during site preparation/staging and pre-treatment of soil. The most common activity describing a BMP used in the first phase of site preparation/staging appears to be the installation of a suitable environmental monitoring system before the commencement of any remediation work at the site. In most cases, this type of BMP may only involve the installation of an air quality monitoring station at the perimeter of the site to address concerns of air pollution in the neighboring community due to site remediation in the future. In some cases, however, the perimeter air quality monitoring may be supplemented by area and/or personal monitoring at the location of soil excavation and treatment. Air quality monitoring may also be supported by monitoring the changes in weather conditions, as shown by one of the case studies. At this same site, it was found that both surface water quality and air quality were being monitored due to concerns about transfer of pollutants in these media during site preparation and pre-treatment of soil.

In addition to the use of environmental monitoring as a BMP, it was reported that site preparation activities often included the construction of new facilities like soil treatment pads (or improvement of existing facilities) that have proved to be effective BMPs (USEPA, 1997). At least two sites were found to have used additional site characterization as a BMP during site preparation/staging. For example, Site No. 1 (Army Ammunition Plant, Minnesota) delineated the actual limits of soil contaminated with lead and other metals before starting any excavation at the site. At the same site, additional studies were conducted to identify and characterize miscellaneous disposal areas existing at the site to enable segregation of the materials found in these areas from the metals contaminated soil. This in-depth characterization of the site reduced the quantity of soil requiring treatment and possibly improved the performance of soil treatment as well. This site also found the establishment of an exclusion zone and decontamination pad for all traffic to/from the site to be effective measures in preventing the cross-media transfer of pollutants during soil excavation.

It was also found that the field use of BMPs was widely introduced during the pre-treatment phase of site remediation. Most sites were found to have dust suppression and stormwater run-on/run-off control in place during site remediation. Different methods of suppressing dust were used during soil excavation and transportation, storage of excavated/treated soil in stockpiles, and soil treatment. At some sites, groundwater protection measures (e.g. providing lined and cover roll-offs for soil treatment residuals) were also introduced as BMPs to control the Cross-Media Transfer in the pre-treatment phase of site remediation. BMPs in the pre-treatment phase of site remediation also included the use of special measures, like selecting the right equipment for excavating soils and providing temporary arrangements for truck movement at the site during construction; preventing the generation of additional wastes or transfer of other pollutants.

The case studies did not identify as many BMPs as originally expected during the treatment phase of site remediation. This observation may be due to the fact that records used in the case studies focused more on the site specific details (in contrast with the design features of the technologies selected for soil treatment at the sites). Most of the BMPs used during soil treatment, as identified by the case studies, were related to the proper management of wastewater and other process intermediates. Site No. 6 (Chromium Manufacturing Plant, Maryland), however, made at least three (3) important changes in its technology for soil containment: (1) providing a capillary break layer in the multi-media cap; (2) preparing multi-media cap for future development of the site as a commercial and recreational area; and (3) providing standby well-heads for additional future groundwater extraction at the site. This site also conducted a series of tests to check the integrity of slurry wall after construction.

The BMPs identified by the case studies included a few that were introduced in the post-treatment/residuals management phase of site remediation. Most of these BMPs involve proper disposal of treatment residuals, including the sorting, decontamination and pre-treatment of residuals to enable off-site disposal of residuals as non-hazardous waste. Proper restoration of the site was also used in some cases as a good and safe management practice. Environmental monitoring – especially the monitoring of groundwater – after completing soil treatment at the site was also considered to be a BMP and a highly desired management practice for a fully protective environmental cleanup.

REFERENCES

Dutta, S. (1996) *Personal Communication with Paul R. Lear.* OHM Remediation Services Corp, May.

Dutta, S. (2000) *Personal Communication with the Site Restoration Manager of the Chromium Manufacturing Plant.* Baltimore Harbor, Baltimore, MD, July.

Dutta, S. (2002) *Environmental Treatment Technologies for Hazardous and Medical Wastes, Remedial Scope and Efficacy.* Tata McGraw Hill Publishing Company, New Delhi.

USEPA. (1997) *Best Management Practices (BMPs) for Soil Treatment Technologies.* Office of Solid Waste, Washington, DC. EPA-530-R-97-007, May.

USEPA. (2000) *FRTR Cost and Performance Remediation Case Studies and Related Information.* Federal Remediation Technologies Roundtable. EPA-542-C-00-001, June.

Chapter 12

Common activities during cleanup operations

The previous chapters covered the technology-specific issues. The discussions were mainly focused on variation in applications of each treatment technology categories, key features of these technology categories, identification and control of cross-media transfer of contaminants during implementation of the treatment technologies. However, at many contaminated sites, whenever any remediation activities are undertaken, there are some common activities that are conducted to implement the cleanup technologies. These activities will have the potential for generating cross-media contamination. The remedial activities generally fall within one of the following four major remedial stages, regardless of the selected technology.

- Site preparation and staging
- Pretreatment
- Treatment
- Post-treatment/residual management.

These stages are not always discrete and separate from one another. For example, residuals management is often an issue while treatment is on-going. In addition, some technologies may not require or have the same level of activity in all stages. However, for the purposes of this chapter, these four remedial steps are treated separately.

This chapter presents best management practices (BMPs) for addressing those remedial activities that are not unique to the technology selected for treating contaminated soil or solid media at a site, but that still pose a potential threat in terms of cross-media contamination. In other words, these BMPs are likely to have applicability to a wide variety of sites because they are associated with a common remedial activity, such as excavation, rather than a specific technology, such as soil washing. The BMPs are organized according to the remedial stage to which they pertain (e.g. staging and site preparation) and then to the applicable cross-media transfer concern (e.g. fugitive dust). The types of remedial activities that may give rise to each concern (e.g. clearing and grubbing, excavation) are also presented to help in determining the applicability of BMPs to a particular site.

As reflected in this chapter, BMPs most commonly associated with activities performed as part of site preparation and staging, pre-treatment, and post-treatment/residuals management are not technology-specific. BMPs associated with activities that occur primarily in the treatment stage of a remediation are generally technology-specific, and hence they are not addressed here. They have been discussed in Chapters 4–10, which cover the individual technology categories.

DOI: 10.1201/9781003004066-12

12.1 GENERAL CROSS-MEDIA TRANSFER POTENTIALS FOR VARIOUS TREATMENT TECHNOLOGIES

During the implementation of any soil treatment technology, the following steps are generally undertaken: (1) Site preparation and staging, (2) Pretreatment activities, (3) Treatment activities, and (4) Post-treatment activities. The specific cross-media concerns during the actual treatment activities are addressed separately under the relevant technology categories in Chapters 4–10. General cross-media transfer potential for contaminants mostly during the site preparation, pretreatment, and post-treatment activities are identified below:

- All soil treatment technology operations carry the risk of inaccurate site characterization. The material encountered at the remediation site may not be identical to the soils studied in treatability or pilot-scale tests. Additional contaminants may be encountered, and the percentage of the fine-grained fraction may be significantly different from that expected. These factors may lead to a long-term storage or generation of high residual volume, thus increasing the cross-media transfer potential of contaminants.
- During several different activities associated with remedy implementations, including staging and site preparation (e.g., clearing, grubbing); drilling, well installation and trenching operations; mobilization and demobilization of equipment; excavation; transport of materials across the site; and some treatment activities, the chances of fugitive dust emissions due to movement of equipment at the site are high. In addition, these same activities can increase the chances of migration of VOCs, SVOCs, and other potentially hazardous materials into the atmosphere.
- During pretreatment operations such as excavation, storage, sizing, crushing, dewatering, neutralization, blending, and feeding, there is the potential for dust and VOC emissions from the contaminated media.
- Migration of contaminants to uncontaminated areas may occur during mobilization or demobilization of equipment.
- VOC and SVOC emissions tend to increase during the hot and dry weather.
- Leaching of contaminants into surface water can occur from uncovered stockpiles and excavated pits.
- Improper handling and disposal of residues (e.g., sediment/sludge residuals or post-washing wastewater) may allow contaminants to migrate into and pollute uncontaminated areas.
- Post-treatment discharges of wastewater, if improperly managed, can cause migration of contaminants.

Table 12.1 provides a summary of the fractional contributions of various remedial activities to the total volatile contaminant emissions, which are potentially a major source of cross-media contamination during many remedial activities.

12.2 GENERAL BMPs FOR SOILS TREATMENT TECHNOLOGIES

Various control practices to prevent potential cross-media transfer of contaminants during cleanup activities have been identified in Tables 13.1–13.5. Also, proper system design is recommended before the implementation of the remedial treatment to avoid cross-media

Table 12.1 Fractional contributions of remedial activities to total volatile contaminant emissions (USEPA, 1991)

Remedial activity	*Fractional contribution of activities to total volatile contaminant emissions for the entire site remediation process*
Excavation	0.0509
Bucket (loading)	0.0218
Truck filling	0.0905
Transport	0.3051
Dumping	0.5016
Incineration	0.0014
Exposed soil	0.0287
Total	1.0000

transfer problems during different treatment steps. However, general BMP options to control specific cross-media transfer of contaminants for different treatment technologies are furnished below:

12.2.1 Site preparation and staging

The following activities are generally carried out before the movement of any equipment on-site:

- Site inspection; surveying; boundary staking; drilling and trenching; sampling; demarcation of hot spots; and construction of access roads, utility connections, and fencing.

Special attention and care are most essential during site preparation activities so that the contaminated media are not disturbed. In the case of unavoidable circumstances, the contaminated media should be subjected to the least possible disturbance/alteration during site preparation activities. The following BMPs are generally recommended:

- Avoid entering the contaminated area. In unavoidable circumstances, build a temporary decontamination area, which could be later used during the cleanup activities. Any above-ground and underground source of contaminants should be identified and located before starting any treatment of the contaminated media.
- Plug and cover any holes or depressions created during the activities, such as soil and soil-gas sampling, field air permeability testing, demarcation of hot spot, etc. to prevent water intrusion. It would also be appropriate to install relevant signs at the same time so that repeated entry to the site is avoided.
- Collect contaminated drilling mud from any drilling operations in a lined/contained system. This will prevent the contaminants from mixing with the normal surface water runoff from the area and the surrounding natural watercourse.
- Protect the contaminated waste generated during site preparation or further site characterization activities as specified in Chapter 13, Tables 13.1–13.4.
- Site investigation and operational plans should take into account the presence of permeable zones, pre-existing underground sewers and electrical conduits.
- Identify surface drainage and subsurface utility systems.

- Incorporate local watershed management goals and priorities into the surface water management plan for the cleanup activities.

12.2.2 Pretreatment activities

Before beginning the actual treatment process, the following activities are generally undertaken:

- Excavation, transportation, storage, sizing, crushing, dewatering, neutralization, blending, installation of feeding systems for contaminated media, and other similar activities.

During the above activities, measures should be taken to control fugitive dust emissions and to prevent releases of contaminated media to the natural environment. Hence, to prevent cross-media transfer of contaminants, the following BMPs are generally recommended for the above activities:

- Remove any aboveground and underground sources of contaminants, such as storage tanks.
- Prevent any offsite runoff from entering and mixing with on-site contaminated media by building earthen berms or adopting similar other measures, as outlined in Table 13.4
- Capture on-site surface water runoff by diverting it to a controlled depression area or lined pit.
- Capture the off gases, volatiles, dust, etc. from sizing, crushing, and blending activities inside a hood or cover or use other options to control these emissions as listed in Chapter 13.
- Capture the dust and VOC emissions associated with these activities that exceed acceptable regulatory limits and treat the vapor/air. Measures for preventing, collecting, and treating dust and VOC emissions are provided in Tables 13.1 and 13.3.
- Collect the contaminated aqueous stream in a lined/contained system when mixing or dewatering. This will prevent the contaminants from mixing with the normal surface water runoff from the area and the surrounding natural watercourse.
- Manage/dispose contaminated debris in a protective manner to prevent cross-media transfer. Protective management includes debris washing, providing covers, testing, and appropriate disposal (see Section 11.7.2 (B) (b)).
- Use proper safety and care to handle explosive wastes, to prevent any explosion during the treatment process. For conducting safe operations, use some of the measures in Section 11.8 of this book, if applicable. When necessary, follow further details and recommendations on safe handling of explosive wastes from the EPA Handbook (USEPA, 1993).
- Ensure that the corrosion factor has been considered in the technology design for all appropriate pipes, valves, fittings, tanks, and feed systems.
- Limit entry to the active site to avoid unnecessary exposure and related transfer of contaminants.
- Use the temporary decontamination area, described in Section 12.2.1 of this book, to keep the site-related contaminants within the active cleanup area.

- Control fugitive dust emissions during excavation by spraying water to keep the ground moist. No spraying is needed during wet weather or rainfall.
- Consider climatological extremes/high wind, etc. when conducting any of the treatment or associated activities. Use real-time weather data to monitor weather conditions and accordingly control treatment operations. In one of the case studies (see Section 11.7 of this book), an onsite weather station was maintained at a nominal cost and was found to be highly useful in controlling weather-related cross-media transfers (Dutta, 2002).
- Monitor VOC emissions during excavation, blending, and feeding of contaminated soils and use appropriate emission control measures.
- Include proper inspection procedures in the operational plans to look specifically for corrosion and wear.
- Check to ensure that the air pollution control device/s are designed for the corrosive nature of the hot gases that are expected to enter these devices, when used in certain soil treatment technologies.
- Use effective erosion control practice by scheduling construction activities to limit the time of exposure of disturbed segments of the site. This entails directing work to one area of a site, then completing and stabilizing that area before moving on to other areas of the site.

12.2.3 Treatment activities

Treatment activities and relevant BMPs are specifically described for each technology category in Chapters 4–10.

12.2.4 Post-treatment activities/residuals management

During the post-treatment process, the following activities are generally carried out:

a. Vapor (gas) phase
    - Collection or destruction of organics
    - Collection of particulates
    - Removal of acid gases.
b. Solid and liquid phases
    - Treatment or disposal of aqueous wastes
    - Disposal of dusts collected as a result of emission control during materials handling, stabilization, or any other tertiary/post-treatment.

While carrying out the above activities, measures should be employed to prevent the release of contaminated media to the natural environment. The following BMPs are generally recommended for this purpose:

- Remedial plans should be checked to ensure that they account for the anticipated differences in characteristics of the treated soil. This may involve the recombination of the treated soil with uncontaminated soil from the site (or off-site) in order to approximate the original soil characteristics prior to contamination. The anticipated soil characteristics of the treated soil should be verified prior to replacement.

- Treated wastes should be checked for leachability prior to disposal in a landfill or other similar systems. Possibilities of long-term degradation and migration of contaminants to groundwater should be carefully evaluated and checked prior to disposal of stabilized/treated material.
- Contaminated debris, soils, and liquid wastes resulting from excavation and installation of wells should be properly handled, either treated on-site or trucked away for off-site disposal. Berms should be built around the active excavation, storage and treatment areas, if necessary, to prevent migration of contaminated runoff away from the area.
- If solid materials such as granulated carbon filters are used to collect emissions, they should be removed carefully from the emissions system to avoid rupturing them and dissipating the contaminated carbon materials. They should be placed into tightly covered containers until they can be recycled or properly disposed of.
- Carbon beds used for VOC removal from the extracted vapor should be properly managed and disposed of in compliance with existing regulations (e.g., US EPA Subtitle C) and should meet all applicable land disposal standards. If the carbon is regenerated using steam or other means, the residual contaminated liquids should be managed as hazardous wastes, and treated or disposed of in compliance with the applicable regulations.
- Containers that hold residual liquids should be stored where they cannot be disturbed or ruptured by large equipment. This may require the construction of a residuals management unit separate from the treatment and storage areas.
- All dusts or other particulates that are collected during emissions control activities should be tested for contamination levels and handled and disposed of properly.
- Air stripping or other treatment of extracted (contaminated) water/liquids should meet all applicable surface water discharge standards for post-treated water.
- When residual treatment wastes are obtained in the form of pure listed waste/liquids (e.g., condensate from steam regeneration of carbon beds), the recycling/reuse option for such residual waste should be considered.

REFERENCES

Dutta, S. (2002). *Environmental Treatment Technologies for Hazardous and Medical Wastes, Remedial Scope and Efficacy*. Tata McGraw Hill Publishing Company, New Delhi.

USEPA. (1991). *Engineering Bulletin, Control of Air Emissions from Materials Handling During Remediation*. Office of Research and Development, Cincinnati, OH. EPA-540-2-91-023, October.

USEPA. (1993). *Approaches for the Remediation of Federal Facility Sites Contaminated with Explosive or Radioactive Wastes*. Office of Research and Development, Washington, DC. EPA-625-R-93-013, September.

Chapter 13

Monitoring and control of cross-media transfer of contaminants during cleanup activities

This chapter describes some of the technologies and practices that are available to monitor and control releases that might migrate and contaminate other media (referred to as cross-media transfer) during the implementation of treatment technologies for soils and/or solid media. The cost of implementing the recommended BMPs is generally subsumed in the overall treatment technology implementation. No specific incremental cost estimates are available at this time for the application of the recommended BMPs. However, based on the information gathered from a few case studies, a short synopsis on the relative cost of implementing BMPs is provided at the end of this chapter.

The control technologies specified in this chapter should generally be applied under the following conditions:

- When a potential risk of cross-media transfer of contaminants exists in association with the use of a soil treatment technology as identified in Chapters 4–10.
- When recommended in the general best management practices (BMPs) section under common activities in cleanup operations (Chapter 12) or technology-specific BMPs in Chapters 4–10.
- When a safe exposure level for workers is exceeded during cleanup activities, as determined by the Occupational Safety and Health Administration (OSHA), per 29 CFR part 1910.
- When any other site-specific conditions that warrant their application, such as proximity to the site of a populated area or a drinking water source.
- When site condition changes due to weather-related impacts or other unprecedented incidents at or near the site.

13.1 AVAILABLE MONITORING AND CONTROL TECHNOLOGIES

Various technologies to monitor and control cross-media transfer of contaminants through – emissions during material handling or cleanup/remedial activities, migration to surface water and groundwater are covered in this section. Specific technologies are recommended in separate tables for selected activities, such as material handling during staging, site preparation and pretreatment activities, primary treatment activity, surface water, groundwater, wastewater, and leachate treatment. A separate table is provided for field monitoring technologies. These foregoing technologies and information are covered mostly in the form of tables (Dutta, 2002).

DOI: 10.1201/9781003004066-13

A variety of new technologies and systems have been developed since then and are available to help meet the monitoring and control needs for site remediation more effectively. Remote sensing is one such technological development, which has been found to help in monitoring the site remediation activities and timely control of leaks or spills from the waste sites, especially when they are remotely located (Roper and Dutta, 2005). Public involvement in the site remediation process has become an important issue since early 2000. As more and more waste sites were slated for cleanup, public involvement in the selection of remedial treatment, access to the site and planning for remedial activities became a part of the regulatory requirements in many states and federal regulations in the US. The use of remote sensing technologies to compile demography and other watershed data has become a useful and effective tool (Dutta, 2003). The following subsection provides a synopsis on the use of remote sensing technologies in municipal and industrial waste site remediation.

13.1.1 Remote sensing technologies for monitoring and control of site remediation activities

Industrial and scientific advances in airborne and satellite remote sensing systems and data-processing techniques are progressively opening up new technological opportunities to develop an increased capability to accomplish the site remediation needs of the industry. These technologies when combined with Geographic Information Systems (GIS) have significant and unique potential for application to a number of cross-cutting issues pertaining to site remediation activities. Advances in geospatial technologies and data analysis methods provide new opportunities for users to increase productivity, reduce costs, facilitate innovation, and create virtual collaborative environments for addressing the challenges of waste site cleanups with improved risk reduction.

Remote sensing and geospatial sensor developments include a new generation of high-resolution commercial satellites that provide unique levels of accuracy in spatial, spectral, and temporal attributes. In addition to the high-resolution panchromatic imagery, there are a number of other commercial imagery products that are potentially applicable to waste site remediation activities. They include airborne and satellite radar, light detection and ranging (LIDAR), multi-spectral, and hyper-spectral sensors. Part of the challenge is matching the best sensor to the specific application. Visualization and advanced data analysis methods are also important capabilities. Automated change detection within a defined sector is one example of analysis capability that will assist in the detection of unauthorized activities, leaks, or spills in areas beyond a remediation site permit boundary. Many waste sites are located in remote areas that are difficult and expensive to monitor. Remote sensing technologies are oftentimes more precise, less expensive, and economical for such monitoring and control activities.

An example of monitoring and control of a waste site involving a surface mining operation near Centralia, state of Washington, USA is provided here. Surface mining activities were conducted there under approved permits from the Office of Surface Mining and Reclamation (OSMRE) in Denver, Colorado, about 1,330 miles (2,150 km) away from the mine site. The OSMRE conducted site monitoring activities using remote sensing methodology. They noticed that the permittee had disturbed 2.3 acres outside of the permit renewal boundary and 0.9 acres outside the overall permit area of the mine. The findings from the site monitoring using remote sensing methodology were later verified during a physical inspection of the site and the permittee was penalized with fines and citation for conducting proper reclamation to the approximate original contour of the areas disturbed outside the permit boundaries. An imagery of this surface mining site overlain by the permit boundary, permit renewal disturbance boundary, and the life of mine-affected area is shown in Figure 13.1 (Dutta, 2003).

A more recent project involved a remote sensing analysis of the Occoquan Reservoir in Fairfax County, Virginia (Roper, 2019). This was a two-year data collection and analysis project using multiple sources and analytical tools to develop a graphic analysis of the reservoir covering four time periods (fall and spring of 2018 and 2019) in the data collection effort. It covered a 10-mile reach from the 80 feet (24.4 m) high dam. The image collection sources for the two-year study included – USGS National Map Program, US Department of Agriculture (USDA) Farm Service Agency, Geospatial Imagery Center, NASA Earth Science Program, ESRI Image Data Center, and Google Earth. The Fairfax Water Authority in Virginia sponsored the study. Fairfax Water serves drinking water to nearly 2 million people within its service area. The source of their water is the Potomac River and the Occoquan Reservoir. This study identified the areas and locations of stream erosion, sources of sediment transport, agriculture, and other sources of chemical pollution in the watershed using remote sensing methodology. A multispectral imagery off the latest eco-sensor aboard the international space station, courtesy of NASA Earth Science Program revealed agricultural chemicals and

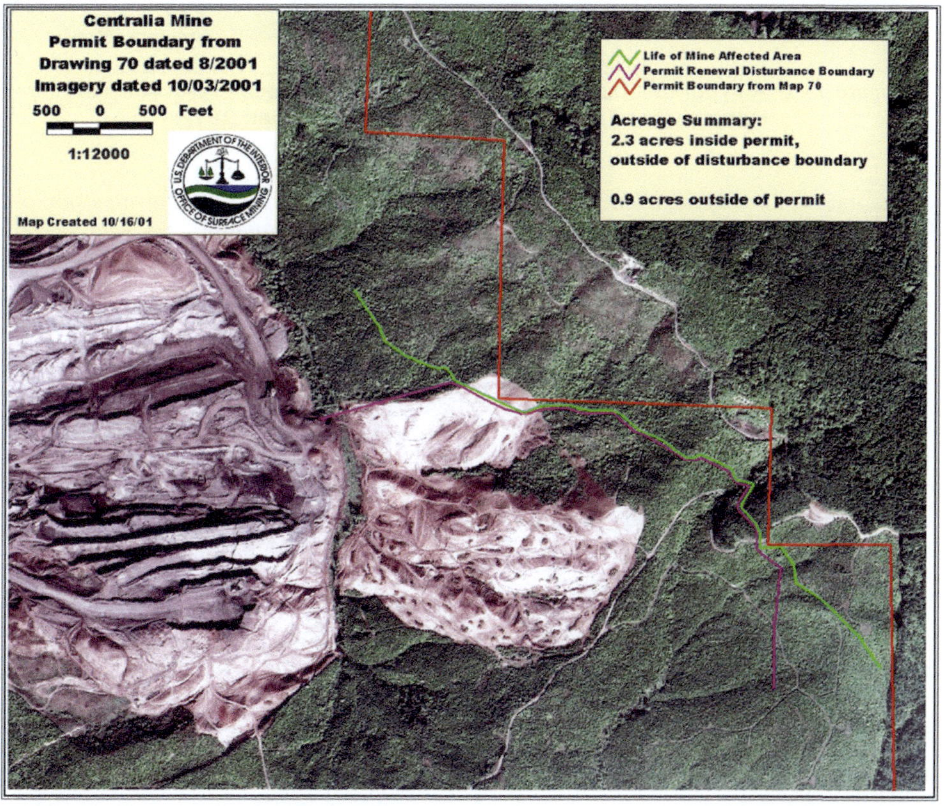

Figure 13.1 An imagery of the Centralia mine site using remote sensing methodology (Dutta, 2003).

Figure 13.2 A hyperspectral imagery of the upper Occoquan reservoir revealing agricultural chemicals and sediment runoff (Roper, 2019).

sedimentation movement from this area as shown in Figure 13.2. The future study plan for this reservoir includes hyperspectral sensor analysis of the reservoir.

The following five tables provide most of the monitoring and control technology information in this chapter (USEPA, 1997). A brief summary of each of these tables is provided below.

13.1.1.1 Emission sources and controls during cleanup activities

Table 13.1 lists potential emission sources that may be encountered during cleanup activities such as containers, tanks, and landfills. It describes some common controls that can be used to reduce those emissions and outlines factors that may contribute to the likelihood of emissions from those sources.

13.1.1.2 Technologies for controlling cross-media transfer of contaminants during materials handling activities

Table 13.2 lists materials handling activities that may be carried out during staging, site preparation, pre- and post-treatment, and the associated activities carrying the risk of cross-media transfer of contaminants. In addition, the table provides information on the control technologies that can be used during those activities as well as the factors that may influence the effectiveness of these control technologies. Some of the controls listed in this table may

Table 13.1 Emissions sources and controls during cleanup activities (Dutta, 2002; USEPA, 1998, 1992a, 1989)

Emission source	*Description of control technologies*	*Factors affecting emissions*
Surface impoundments	*Air-supported Structures* are made of light materials (often plastics, vinyl, or coated fabrics) that form a roof-like structure over the impoundment. Fans are used to maintain positive pressure to inflate the structure. For effective control, the air vented from the structure should be sent to a control device such as a carbon adsorber. Air-supported structures have been used as enclosures for conveyors, open top tanks, storage piles, and impoundments. *Floating Membrane Covers* are used to cover large impoundments containing liquids. The membrane must provide a seal at the edge of the impoundment, and provisions must be made for the removal of rainwater that accumulates on the covers. In addition, vent systems for the removal of accumulated gases and pumping systems for the removal of accumulated sludge may be necessary.	• Volatility of constituent • Residence time • Surface area • Turbulence • Wind speed • Temperature • Extent of competing mechanisms (e.g., biodegradation)
Tanks	*Fixed roofs* can be retrofitted to open tanks, or a fixed-roof tank can be used to replace an open tank or impoundment. Compared to an open tank, a fixed roof tank can provide additional 86%–99% emission control. *Floating roofs* are common on tanks at petroleum refineries. The roof floats on the liquid and moves with changes in the liquid level, controlling working losses. Floating roofs can be installed internally in a fixed-roof tank or externally in a tank without a fixed roof. Emissions from a properly maintained floating roof are very low.	• Volatility of constituent(s) • Surface area • Turbulence • Wind speed • Temperature
Dewatering Devices	Dewatering devices, such as rotary drums and presses, provide several opportunities for volatile organics to be emitted, such as when a press is opened to remove and transport accumulated sludge, or during pressing, when volatile liquids may leak from a press into a drip pan underneath. Emissions from dewatering devices can be controlled by building an enclosure around the unit and venting it to a control device (best used for presses or rotary devices) or by collecting volatile organics in a condenser above the volatile source, treating the waste, and discharging it as appropriate (best for thin-film evaporators). In addition, sludge fixation often generates volatiles during mixing, when agitation is provided while adding the fixative agent. Emissions during fixation can be controlled by installing covers or enclosures that are vented to a control device.	• Temperature • Surface area • Turbulence • Wind speed • Concentration • Volatility
Containers	*Submerged fill piping* has been shown to decrease emissions by 65% relative to splash filling. In submerged filling, an influent pipe is inserted below the existing liquid surface in the container. Liquid is introduced into liquid, rather than spilled on top of the liquid surface, which reduces splashing and the degree of saturation of the displaced vapors.	• Volatility of constituent(s) • Surface area • Turbulence • Wind speed • Temperature
Landfills	*Carbon Adsorption, Condensation, Absorption, or Vapor Combustion* are traditionally used to capture and control emissions.	See Table 13.3 for descriptions of air control technologies

Table 13.2 Technologies for controlling the cross-media transfer of contaminants during materials handling activities (USEPA, 1998, 1991, 1988)

Remedial activities	*Description of control technologies*	*Controlling factors*
All	*Operational controls*: Those procedures or practices inherent to most site remediation projects that can be instituted to reduce VOC and particulate matter emissions. Use the following guidelines to the extent possible: • Plan site remediation for times of year with relatively cooler temperatures and lower wind speeds to minimize volatilization and particulate matter emissions. • Maintain lower speeds with all vehicles on unpaved roads. • Control placement and shape of storage piles. Place piles in areas shielded from prevailing winds. Shape pile in a way that minimizes surface area exposed to wind. • During excavation, use larger equipment to minimize surface area/volume ratio of material being excavated. • During dumping, minimize soil drop height onto pile, and load/unload material on the leeward side of pile. • During transport, cover or enclose trucks transporting soils, increase freeboard requirements, and repair trucks exhibiting spillage due to leaks.	• Temperature • Moving equipment or vehicle speed • Shape and location of storage piles • Equipment size • Covering waste or excavated soil during Transport. • Leak-free containers.
Excavation	*Covers and physical barriers*: Physically isolate the contaminated media from the atmosphere. Include soils (topsoil or clays); organic solids such as mulch, wood chips, sawdust, or straw, typically anchored with a net; asphalt/concrete; gravel/slag with road carpet; synthetic covers (e.g., tarps). Some technologies best used in active areas, others in inactive areas (USEPA, 1992b).	Site characteristics (terrain, vegetation, nature of contaminated media) and access needs.
	Foam coverings: "Blanket" the emitting source with foam, thus forming a physical barrier to emissions. Also insulate emitting source from wind and sun, further reducing particulate and volatile emissions. Several commercially available. Generally used in active areas.	Drainage rates, wind speed, precipitation, surface roughness, temperature, surface activity, contaminant characteristics.
	Wind screens: Provide an area of reduced velocity that allows settling of large particles and reduces particle movement from exposed surfaces on leeward side of screen. Also reduce soil moisture loss due to wind, resulting in decreased VOC and particulate emissions.	Windscreen porosity, wind direction with respect to screen, windscreen height, soil silt content.
	Slurry cover sprays: Spray soil piles/excavated areas with a thin layer composed of a fibrous slurried aggregate that hardens to form a protective layer. A special agent is used to prepare the fibrous slurry by mixing two products with water just before use (Dutta, 2002).	

(*Continued*)

Table 13.2 (Continued) Technologies for controlling the cross-media transfer of contaminants during materials handling activities (USEPA, 1998, 1991, 1988)

Remedial activities	*Description of control technologies*	*Controlling factors*
	Water sprays: Agglomerate small particles with larger particles or with water droplets. Also, water added to the soil cools the surface soil and decreases air-filled soil porosity, both of which reduce VOC emissions.	Application rate, application frequency, meteorological conditions, traffic rate.
	Water sprays with additives: Common additives include hygroscopic salts, bitumens, adhesives, and surfactants. Reduce emissions by adsorbing moisture from the air, thereby increasing the soil moisture content; agglomerating surface soil particles to form a surface crust; or reducing water surface tension, thereby increasing the wetting capacity of the water.	
	Enclosures: Usually self-supported or air-supported structures; for soil storage piles, usually self-supported structures similar to the "beehive" used to store road salts. Provide a physical barrier between the emitting area and the atmosphere.	Potential for enclosure materials to react with contaminants.
Transportation	*Covers and physical barriers*: Road carpets are water-permeable polyester fabrics that are placed between the roadbed and a coarse aggregate road ballast, such as gravel, across which vehicles travel. Creates a physical barrier between moving vehicles and source of emissions. *Covers on loads*: Cover all loads being moved by truck, open piping, or other conveyance with tarps, roofs, or other structures that will eliminate or reduce the likelihood of particulate release into the atmosphere. *Water sprays of active areas*: See Excavation *Dust suppressants*: See Excavation	
Dumping	*Water sprays*: Water can be sprayed in a curtain-like fashion over the bed of a truck (or over any conveyance system, such as a moving belt) during dumping; see Excavation for details on how water sprays work *Water sprays with additives*: Use like water sprays (see above), with additional substances such as surfactants; see Excavation.	
Preparation and Feeding of Contaminated Media and Feeding Media into Remediation System	*Covers and physical barriers*. See Excavation *Enclosures*. See Excavation *Collection hoods*: Commonly used in small areas (e.g., waste stabilization/ solidification mixing silos, bioremediation reactors) and route those emissions to air pollution control devices. Capture emissions by creation of an airflow after the emitting source that is sufficient to remove the contaminated air.	Distance between hood and emissions source; volumetric flow rate into hood; surrounding air turbulence; hood design.
Storage of Waste/Residuals	*Covers and physical barriers*: See Excavation *Foam coverings*: See Excavation *Wind screens*: See Excavation *Water sprays*: See Excavation *Water sprays with additives*: See Excavation	

be applicable to the treatment activities, which are discussed in the individual remediation technology in Chapters 4–11.

13.1.1.3 Technologies for reducing contaminant concentrations in air emissions generated during cleanup/remedial activities

Table 13.3 has the control technologies that can be used to reduce the concentrations of contaminants in air emissions. This table describes each technology and outlines factors that may influence the effectiveness of those control technologies.

13.1.1.4 Technologies for controlling cross-media transfer of contaminants to water

Table 13.4 provides a list of controls that should be considered during all remedial activities to minimize the potential for releases from soil to surface water and/or groundwater. The examples provided are for relatively small-scale structures that can be applied to short-term projects. For larger-scale and long-term projects, documents (MWCOG 1992 and others) listed under references at the end of this chapter should be consulted.

13.1.1.5 Field monitoring technologies

Table 13.5 provides a list of technologies or practices that can be used to monitor potential emissions during remediation activities. It describes the technologies that can be used to monitor emissions from active and inactive sites. These technologies can be applied before and during remediation, as needed. A few simple and easy-to-use monitoring techniques are also listed in this table.

13.2 RELATIVE COSTS OF IMPLEMENTING, MONITORING AND CONTROL TECHNOLOGIES (BMPs)

A screening-level analysis of the costs associated with implementing the practices outlined in this chapter was conducted. The cost analysis of implementing the BMPs is generally complicated due to the following factors:

- Many of the suggested practices do not involve equipment purchases nor do they entail a well-defined or discrete task outside the integral remediation activities.
- Cost data available on soil cleanups tend to be aggregated such that costs for performing specific practices are not separable from the overall cost of treatment technology implementation. In addition, the practices applicable or necessary for a particular cleanup, and the magnitude to which they are performed, can vary based on the characteristics of the site and of the contaminants present.

As a general rule, no considerable change/increase in cost is expected for implementing the monitoring and control practices suggested in this chapter. At many sites in the US, the remedial project managers are already implementing these controls as good management practices whenever performing cleanups. The costs of implementing these control practices

Table 13.3 Technologies for reducing contaminant concentrations in air emissions generated during remediation

Technology	*Description*	*Factors influencing effectiveness*
Incineration		
• Catalytic (also known as catalytic oxidation)	Contaminant-laden waste gas is heated with auxiliary fuel to between 600°F and 900°F. The waste gas is then passed across a catalyst where the VOC contaminants react with oxygen to form carbon dioxide and water (USEPA, 1992b).	Waste gas composition
• Regenerative thermal	Contaminated air is preheated, then combusted to oxidize the organic volatiles. The clean gas exiting the combustion chamber is cooled by passing through cool packed beds, then discharged to the atmosphere. The remaining contaminated air is reheated, then passed through packed beds with clean air, cooled and discharged. The cycle of heating, cooling, and discharge is repeated as necessary (USEPA, 1995).	Useful for low concentrations of VOCs at low to moderate feed rates
Adsorption	Organics are selectively collected on the surface of a porous solid. Typical adsorbents include activated carbon, silica and aluminum-based adsorbates (USEPA, 1995).	Must be used in conjunction with units that recover or destroy organic volatiles
• Non-regenerable	Air stream containing volatiles flows upward through one or two fixed beds of adsorbent. Volatiles are adsorbed until breakthrough occurs, at which time adsorbent is replaced (USEPA, 1995).	At high-influent concentrations, not generally cost-effective because of large volumes of adsorbent that must be used
• Modified	Systems designed specifically for low concentrations (less than 100 ppm) of organic volatiles in gas, which most treatment systems are not designed to accommodate. Adsorption treatment is followed by treatment of the concentrated volatiles in the regenerated gas (by incineration), or volatiles below regulatory limits are discharged into the atmosphere (USEPA, 1995).	
• Fabric filter	Designed for the control of particulate emissions from point sources. One or more isolated compartments containing rows of fabric bags or tubes. Particle-laden gas passes up along the surface of the bags then radially through the fabric. Particles are retained on the upstream face of the bags, while the clean gas stream is vented to the atmosphere (USEPA, 1992b).	Flue gas temperature, gas stream composition, particle characteristics.

(Continued)

Table 13.3 (Continued) Technologies for reducing contaminant concentrations in air emissions generated during remediation

Technology	*Description*	*Factors influencing effectiveness*
• High efficiency particulate air filter (HEPA)	Used at sites requiring 99.9% or greater particulate removal. Can be used as a particulate matter polishing step in ventilation systems for enclosures or with solidification/stabilization mixing bins (USEPA, 1992b). Comprised of a series of filters, filter housing, duct work, and a fan. Filters are aligned in series, in parallel, or in a combination. Air is forced over the filters; larger particulates are collected on prefilters, finest particulates are collected on filters. When breakthrough occurs, filters are replaced and disposed of.	Moisture content of contaminated air stream; degree of particulate matter loading.
Absorption	Organics in the gas stream are dissolved in a solvent liquid, such as water, mineral oil, or other nonvolatile petroleum oil. The contact between the absorbing liquid and the vent gas is accomplished in a counter-current flow. A few common flow media are listed below.	Works better for highly volatile compounds with higher volatile concentrations.
• Spray towers • Packed columns • Scrubbers	The solvent liquid flows from the top of the spray towers, while the vent gas containing contaminants are pushed up in a counter-current flow through a granular/porous media in – packed tower, or packed columns or scrubbers In most cases, the volatiles are absorbed in the solvent liquid; the volatile compounds are then recovered from the solvent by a condenser as liquids (USEPA, 1995).	
Other commercial technologies		
• Enhanced Adsorption	Combines wet scrubbing, carbon adsorption, and ozone reactions; ultimately, all organic volatiles are oxidized to carbon dioxide, water, and hydrochloric acid (if chlorine is present in the contaminated air stream) (USEPA, 1995).	Periodic replacement/ regeneration of saturated filter media provides smooth and effective operation.
• Internal Combustion Engines	Uses a conventional automobile or truck internal combustion engine as a thermal incinerator of contaminated gas streams (USEPA, 1992b).	Optimum air/fuel mixture for complete combustion.
• Membranes	Membrane concentrates organic solvents by being more permeable to organic constituents than to air. A pressure difference is imposed across a selective membrane (with a compressor or vacuum), which drives the separation of the solvent from the gas stream. The stripped-off gas is either vented or recycled to the source of contamination (USEPA, 1992b).	Solvent permeability (flux across the membrane), separation factor (degree of concentration the membrane can achieve)
• Condenser	Volatile components of a vapor mixture are separated from the remaining gas by a phase change. Condensation occurs when the partial pressure of the volatile components is greater than or equal to its vapor pressure, which can be achieved by lowering the temperature or increasing the pressure of the gas stream (USEPA, 1992b).	Characteristics of vapor stream, condenser operating parameters

(Continued)

Table 13.3 (Continued) Technologies for reducing contaminant concentrations in air emissions generated during remediation

Technology	*Description*	*Factors influencing effectiveness*
• Wind screens	Provide limited control of VOC emissions by increasing the thickness of the laminar film layer (stagnant boundary layer) on the leeward side of the screen; also reduce soil moisture loss to wind, resulting in decreased VOC emissions (USEPA, 1992b).	Wind screen porosity, wind direction with respect to wind screen, height of the screen, soil silt content
Emerging technologies		
• Corona discharge	Uses a high voltage/low current electrical charge to destroy a wide range of molecules in a gas stream containing organic volatiles (USEPA, 1995).	
• Heterogeneous Photocatalysis	Uses a near-ultraviolet light to continuously activate a semiconductor (such as titanium dioxide). The activated surface of the semiconductor then acts as a catalyst for the oxidation of the organic volatiles in the air (USEPA, 1995).	Possibly contaminant concentrations (incomplete reactions when concentrations are high); humidity (high humidity may reduce effectiveness)
• Biofiltration	Off-gases containing biodegradable organic compounds are vented, under controlled temperature and humidity, through a biologically active material. The microorganisms contained in the bed of compost-like material digest or biodegrade the organics to carbon dioxide and water (USEPA, 1995).	
• Electrostatic Precipitators	Electrostatic precipitators (ESPs) are advanced air pollution control devices which can be dry or wet – wire pipe or wire plate types. Dry wire-pipe ESPs are occasionally used by the textile industry, pulp and paper facilities, the metallurgical industry, including coke ovens, hazardous waste incinerators, and sulfuric acid manufacturing plants, among others, though other ESP types are employed as well. Wet wire-pipe ESPs are used much more frequently than dry wire-pipe ESPs, which are used only in cases where wet cleaning is undesirable, such as high-temperature streams or wastewater restrictions. The design efficiencies of typical new ESPs are between 99.0% and 99.9% (USEPA, 2003a).	High-temperature stream and wastewater restrictions may exclude the use of wet ESPs. Pollutant loading with large particle size may require pre-treatment, such as mechanical collectors to reduce the load on the ESP.
• Ultra low penetration air (ULPA) filter and high efficiency particulate air (HEPA) filters	These filters have collection efficiencies of 99.99% for removing particulate matters (PM) with 0.12 μm in diameter or larger. These filters can be effectively used for removing pollutants with PM greater than or equal to 0.3 micrometer (μm) in diameter, and PM greater than or equal to 0.12 μm in aerodynamic diameter that is chemically, biologically, or radioactively toxic. It can also remove hazardous air pollutants (HAPs) that are in particulate form, such as most metals (except mercury) (USEPA, 2003b).	These filters cannot trap emissions that are in the form of elemental vapors, such as mercury.

Table 13.4 Technologies for controlling cross-media transfer of contaminants to water

Technologies	*Descriptions*	*Purpose*
Temporary Diversion	A temporary ridge or excavated channel, a combination of ridge and channel, constructed across sloping land on a predetermined grade (USDA, 1995).	Protects work areas from upslope runoff and diverts sediment-laden water to an appropriate sediment trapping facility or stabilized outlet.
Filter Berms	A temporary ridge of gravel or crushed rock constructed across a graded right-of-way (USEPA, 1972).	Retains sediment on-site by retarding and filtering runoff while at the same time allowing construction traffic to proceed along the right-of-way. Used primarily across graded rights-of-way that are subject to vehicular traffic. Also applicable for use in drainage ditches prior to roadway paving and establishment of permanent ground cover.
Infiltration basins	Impoundments where incoming stormwater runoff is stored until it gradually exfiltrates through the soil of the basin floor. Pollutant removal is accomplished by adsorption, straining, and microbial decomposition in the basin subsoils as well as the trapping of particulate matter within pre-treatment areas (MWCOG, 1992).	Collects sediments and pollutants
Temporary sediment traps	Small, temporary ponding basins formed by the construction of an embankment or excavated basin (USDA, 1995).	Detains sediment-laden runoff from small disturbed areas for a sufficient period of time to allow the majority of sediment and other floating debris to settle.
Diversion dikes	A combination of ridge and excavated channel constructed to divert surface flow	Diverts overland flow away from any unstabilized or contaminated areas (USDA, 1995).
Riprap	A combination of large stones, cobbles, and boulders used to line channels, stabilize banks, reduce runoff velocities, or filter out sediments (MWCOG, 1992).	Prevents erosion on steep or cleared slopes (USDA, 1995).
Sand filters	A filtration system constructed of layers of peat, limestone, and/or topsoil, which may also have a grass cover crop. The first flush of runoff is diverted into a self-contained bed of sand. The runoff is then strained through the sand, collected in underground pipes, and returned to the stream or channel (MWCOG, 1992).	Treats stormwater runoff; removal rates for sediments and trace metals are high, and moderate for nutrients, BOD, and coliform
Silt fence	A fencing made of woven geo-textile filter fabric, placed across a sloping disturbed area, typically 24 inches (60 cm) high to trap sediment and small debris from runoff (USDA, 2014).	Traps sediment and small debris.

Table 13.5 Field monitoring technologies (Freeman and Harris, 1995; USEPA, 1994)

Techniques	*Descriptions of application to air emissions*	*Applicable to*	*Types of detectors commonly used*
Direct measurement with hand-held equipment	Hand-held organic vapor analyzers provide quick readings on presence of organic vapors. These analyzers can be used to check for emissions from specific equipment (e.g., pipe seals, gaskets), or to identify when emission levels change from one area to another.	Volatile organic compounds (VOCs), some semi-volatile organic compounds (SVOCs)	Organic Vapor Analyzer (OVA)
Headspace analysis	Involves collecting waste material in a bottle with "significant" headspace and allowing the waste/headspace to reach equilibrium. The headspace gas is then analyzed for volatile compounds with simple real-time analyzers.	VOCs, SVOCs	OVA, photoionization detector (PID) for VOCs and SVOCs
Real-time instrument survey	Screening takes place directly over the waste to obviate modeling by testing the air above the surface. This approach can identify "hot spots" of emissions and zones of similar emissions[a]	VOCs, SVOCs, Particulate Matters (PM)	OVA, PID for VOCs and SVOCs; dust monitor (DM) for PM
Upwind/ downwind survey	Monitors upwind/downwind concentrations of ambient target compounds. Often, real-time analyzers with flame ionization and photoionization are used for detecting organic emissions.	VOCs, SVOCs	OVA, PID for VOCs and SVOCs; DM for PM; gas chromatography/ mass spectrometry (GC/MS)
Surface flux chamber	A direct measurement approach applicable to many kinds of waste sites and capable of generating both undisturbed and disturbed emission rate data for volatile and semivolatile compounds. The technology uses a chamber to isolate a surface-emitting gas species (organic or inorganic); emission rates are calculated by measuring the gas concentration in the chamber and using the chamber sweep airflow rate and surface area.	VOCs, SVOCs, PM	OVA, PID for VOCs and SVOCs; specific compound detector (SD), GC/MS
Transect	An indirect method that involves the collection of ambient concentration data for gaseous compounds and/or particulate matter using a two-dimensional array of point samplers. These data, along with micro-meteorological data, can be used to estimate the emission rate of the source by using a dispersion model. Data that represent emissions from a complex or heterogeneous site or an activity that generates fugitive air emissions can be obtained.	VOCs, SVOCs	Fourier Transform Infrared Optical Remote Sensing Detector; Ultraviolet-Differential Optical Absorbance Sensor; Filter Band Pass Absorption Detector, Laser, Perimeter Air Sampling (PAS)

(*Continued*)

Table 13.5 (Continued) Field monitoring technologies (Freeman and Harris, 1995; USEPA, 1994)

Techniques	*Descriptions of application to air emissions*	*Applicable to*	*Types of detectors commonly used*
Visual inspection	Periodic visual inspection of pipes and joints for corrosion and leaks could provide early detection and prevent major leaks or spills.	Liquids and gases	Occasionally aided by hand-held telescopes or magnifying glass.
Periodic watershed evaluation	The impact of cleanup activities on the watershed could be periodically evaluated by monitoring the following indicator parameters: • Level of siltation • dumping of materials, tools, or equipment related to cleanup activities • Water clarity • Habitat and vegetation. Most of the above monitoring could be accomplished by visual inspections.	Water, sediment, and other related materials.	Evaluation and analysis of visual findings.

[a] For details on screening survey, monitoring instruments, limitations of portable VOC detection devices, performance criteria of VOC detectors, data handling, and calibration procedures, see cited reference – USEPA (1994, pp. 37–47).

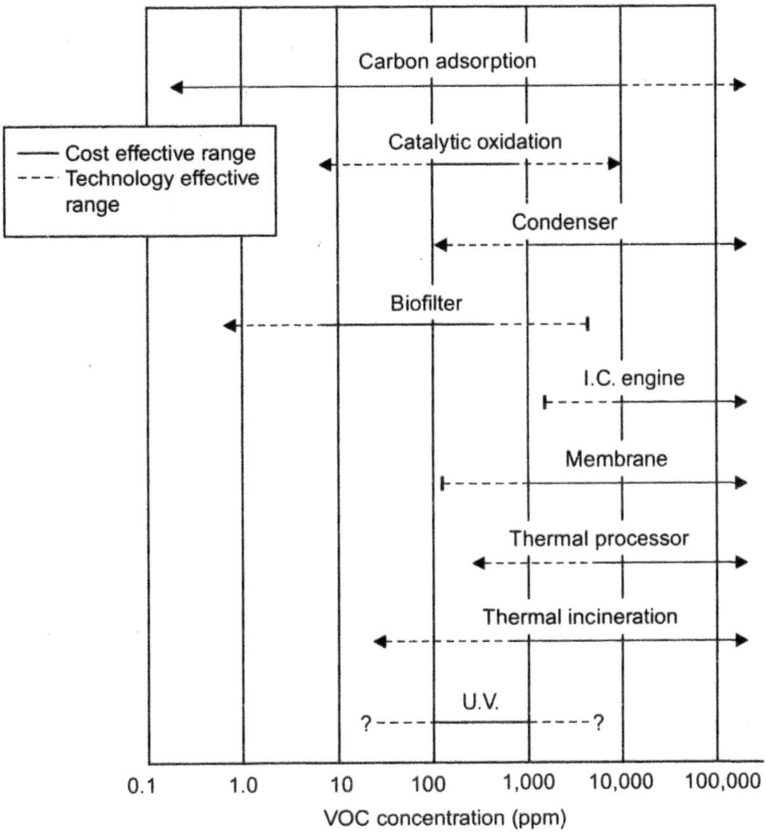

Figure 13.3 Relative cost effectiveness for point source VOC controls (USEPA, 1992b).

are subsumed in the overall cleanup costs. In some cases, additional costs might have to be incurred to implement cross-media transfer prevention controls. However, in such cases, the costs would probably have to be incurred to meet the existing state or federal cleanup requirements or to avert potential future costs to address cross-media transfers of contaminants. Based on preliminary cost data gathered during field validation (USEPA, 1997), these costs are generally an integral part of the total cost of the remedial activity and are estimated to be two to six percent (2%–6%) of the overall cleanup cost.

Although actual cost figures are not available at this time for the recommended BMPs, studies have been conducted earlier by EPA on the relative cost effectiveness for point source VOC controls as shown in Figure 13.3 (USEPA, 1992b).

REFERENCES

Dutta, S. (2002) *Environmental Treatment Technologies for Hazardous and Medical Wastes, Remedial Scope and Efficacy*. Tata McGraw Hill Publishing Company, New Delhi.

Dutta, S. (2003) A Model Watershed Management Plan With Stakeholder Partnership. *ESRI User Conference paper no. 114*, San Diego, CA, July.

Freeman, H. M., and E. F. Harris. (1995) *Hazardous Waste Remediation: Innovative Treatment Technologies*. Technomic Publishing Co., Inc., Lancaster, PA.

Metropolitan Washington Council of Government (MWCOG). (1992) A Current Assessment of Urban Best Management Practices, a report prepared for U.S. EPA's Office of Wetlands, Oceans, and Watersheds, March.

Roper, W. E. (2019) *Remote Sensing Analysis of the Occoquan Reservoir, A Water Quality Study Presentation*. Fairfax Water Authority, Fairfax, VA, December.

Roper, W. E., and S. Dutta. (2005) Remote Sensing and GIS Applications for Pipeline Security Assessment. *ESRI User Conference paper no. 1762*, San Diego, CA, July.

USDA. (1995) Illinois Urban Manual – A Technical Manual Designed for Urban Ecosystem Protection and Enhancement, prepared for Illinois Environmental Protection Agency, by the U.S. Department of Agriculture (USDA), Natural Resources Conservation Service, Champagne, IL.

USDA. (2014) *Temporary Pollution Control – Silt Fence*. USDA, Natural Resources Conservation Service.

USEPA. (1972) *Guidelines for Erosion and Sediment Control Planning and Implementation*. EPA-R2-72-015, Office of Research and Development, Washington, DC, August.

USEPA. (1988) *Project Summary. Fugitive Dust Control Techniques at Hazardous Waste Sites: Results of Three Sampling Studies to Determine Control Effectiveness*. EPA/540/S2-85/003.

USEPA. (1989) *Seminar Publication, Corrective Action Technologies and Applications*. Office of Research and Development, Cincinnati, OH. EPA/625/4-89/020.

USEPA. (1991) *Engineering Bulletin, Control of Air Emission from Materials Handling During Remediation*. Office of Research and Development, Cincinnati, OH. EPA-540-2-91-023, October.

USEPA. (1992a) *Seminar Publication, Organic Air Emission from Waste Management Facilities*. Office of Air Quality Planning and Standards, Research Triangle Park, NC. EPA/625/R-92/003, August.

USEPA. (1992b) *Control of Air Emissions from Superfund Sites*. Office of Research and Development, Cincinnati, OH. EPA-625-R-92-012, November.

USEPA. (1994) *Control Technologies for Fugitive VOC Emissions from Chemical Process Facilities, Handbook*. EPA/625/R-93/005, Office of Research and Development, Cincinnati, OH, March.

USEPA. (1995) *Survey of Control Technologies for Low Concentration Organic Vapor Gas Streams*. Office of Air Quality Planning and Standards, Research Triangle Park. EPA/456/R-95/003, May.

USEPA. (1997) *Best Management Practices (BMPs) for Soil Treatment Technologies*. Office of Solid Waste, Washington, DC. EPA-530-R-97-007, May.

USEPA. (1998) *Stationary Source Control Techniques Document for Fine Particulate Matter*. Office of Air Quality, Research Triangle Park, NC. EPA-452-R97-001, October

USEPA. (2003a) *Air pollution Control Technology Fact Sheet*. EPA, Research Triangle Park, NC. EPA-452-F-03-027.

USEPA. (2003b) *Air pollution Control Technology Fact Sheet*, EPA, Research Triangle Park, NC. EPA-452-F-03-023.

Chapter 14

Treatment options for medical waste

This chapter presents a synopsis of medical waste management and provides some details on available options for the treatment of medical waste. The author hopes that this information would meet the needs of interested readers from different parts of the globe. In the current pandemic situation involving SARS-CoV-2 – the virus that causes COVID-19 – how some of the countries, such as Spain and South Korea, have dealt with the unprecedented surge of waste from health-care facilities is presented briefly in this chapter. In addition, guidance and information related to the safe management of medical waste from infectious diseases and other normal hospital/nursing home wastes, recommended by various agencies are incorporated in this chapter (CDC, 2020; WHO, 2014, 2018, 2020).

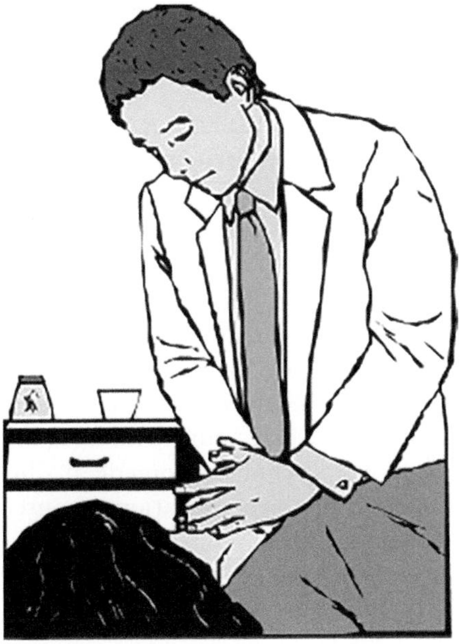

The environmental issues associated with the generation, management, and disposal of waste from medical facilities are quite extensive and are beyond the scope of this book. However, the synopsis provides a general profile, and the detailed treatment options for medical waste in this chapter are expected to cover similar issues and concerns in many parts of the world.

DOI: 10.1201/9781003004066-14

The main focus of this chapter is on the environmentally safe management of medical waste in developing and the developed countries. Information provided here should help identify the necessary tools and techniques for safe handling and management of medical waste.

There are many developing countries where hospital wastes as a whole are being dumped with municipal garbage (Pathak, 1998). In some other countries, medical waste has not even been legally defined (Dutta, 1998; Kwok-Kuen, 1998). Many countries are in the process of developing plans and regulations for the safe management of medical waste. Sample questions and issues often raised by professionals from developing countries such as Bangladesh, India, Malaysia, Nepal, and others include:

- There is no specific 'definition' of hospital wastes.
- What should be done with no regulations in place.
- Rules are framed, but we are still in the dark as they are not well defined.
- Management of medical waste is not given priority, although public interest is now building up.
- No cost–benefit analysis is available for any medical waste treatment technology, other than the incinerator. Awareness is low, and technology is non-existent.
- No training/awareness for hospital staff at any level.

Incidents of public exposure to discarded blood vials, needles, empty bottles, and syringes from municipal garbage bins and disposal sites are not uncommon in countries where medical waste is not properly managed. Thus, in most developing countries, there is an urgent need for environmentally safe management of medical waste. Given the prevailing scenario, the author suggests the slogan:

Segregation, Segregation, and Segregation

as one of the key steps towards a safer, smarter, better, and cheaper management of medical waste.

Although the term "medical waste" often connotes potentially hazardous or infectious materials, such as dressings from wounds or needles and syringes, the majority of the waste generated by healthcare providers results from administrative and housekeeping activities and resembles the waste from our homes or offices. Thus, a key aspect of responsible waste management is to distinguish and *segregate* this basic solid waste fraction from wastes of greater concern, including biomedical, chemical, chemotherapeutic, and low-level radioactive wastes. Effective segregation at source involves mutilation of sharps and plastics before disposing of these items to prevent their reuse and avoid the spread of any infections from possible reuse. A needle destroyer or hub cutter is generally used to destroy or mutilate the needles, which can be placed in tamper-proof and leak-proof containers for disposal.

One challenge facing health-care professionals, government agencies, and environmentalists is the establishment of procedures to reduce the total quantity of medical waste generated. This reduction can be achieved most effectively by *segregating* the solid wastes from the hazard-bearing wastes at the source. It may be enhanced by integrating waste prevention concepts into procurement procedures and training for medical personnel, including doctors, nurses, and laboratory technicians.

Waste reduction can be optimized through a comprehensive program to oversee its environmentally sound *Segregation* technique such as source separation, storage, transportation, treatment, and disposal. By providing training for staff whose responsibilities include materials handling and waste management, healthcare administrators can help reduce the impact of medical waste on human health and the environment. The successful implementation of a medical waste management program requires significant cooperation among the responsible parties and commitment in terms of time and resources.

Medical facilities can adopt various approaches to reduce the environmental impact of medical waste. The appropriateness and effectiveness of these approaches will, however, depend upon the local regulations and priorities. The basic framework of a medical waste reduction and management program constitutes many variables and the effectiveness of a new approach may be influenced substantially by the environmental quality concerns at the national, regional, and local levels. Another key issue is the ability and willingness of the responsible parties to review their current practices and adopt prevention of waste generation as a priority. Limited access to financial resources as well as to scientific and technical assistance may also limit program success or the program may stop short of achieving technological sophistication. However, it is possible for creative and flexible administrators and organizations to implement innovative waste prevention and management procedures – even without abundant financing.

14.1 COMPOSITION OF MEDICAL WASTE

The term "medical waste" is applied to a very broad range of materials that includes all of the solid wastes generated by medical service providers in the course of the diagnosis, treatment, and/or prevention of diseases, including medical research. Medical waste can be divided into two broad categories: municipal solid wastes and wastes that may create public health or environmental risks if inappropriately managed or disposed of (USEPA, 1990).

Most of the wastes from medical activities include significant amounts of the solid wastes, similar to what is generated by other businesses and residences. As shown in Figure 14.1, a significant proportion (about 85%) of all waste from health-care facilities is non-hazardous waste and is usually similar in characteristics to municipal solid waste. More than half of all non-hazardous waste from hospitals is paper, cardboard, and plastics, while the rest comprises discarded food, metals, glass, textiles, plastics, and wood. About 10% of the waste generated during normal condition are infectious, and 5% of the waste contain hazardous chemicals or radioactive elements. These medical wastes, however, require special precautions because of the risk of transmitting disease or of hazards from exposure to chemical hazards or radioactivity. The fraction containing infectious wastes also has pathological wastes, and sharps, such as used needles or scalpel blades. It may also include small quantities of hazardous materials, such as discarded pharmaceuticals, cleaning products, and chemical solvents. In addition, the nuclear medicine departments may generate small amounts of low-level radioactive wastes from diagnostic procedures which become a part of the chemical/radioactive waste (WHO, 2014). Figure 14.2 shows a typical waste storage area of a medical facility in Asia.

Figures 14.3 and 14.4 provide a comparative illustration of the fractional components of medical waste, and the solid waste stream, in hospitals in the US and India. The common solid wastes from these hospitals consist of paper from offices; corrugated cardboard from deliveries of supplies and equipment; glass, metal, paperboard, and plastic packaging; and

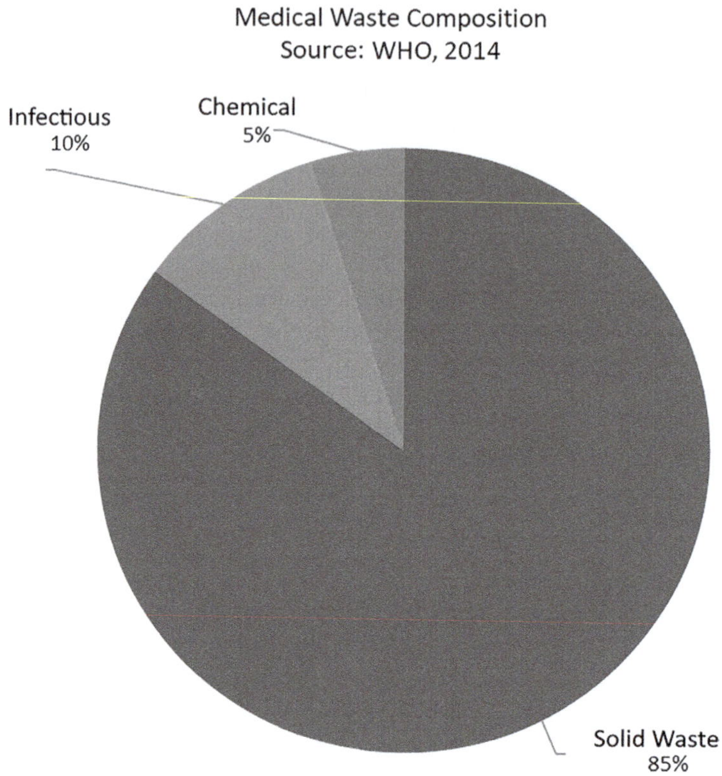

Figure 14.1 Medical waste composition.

Figure 14.2 Typical waste storage of a medical facility in Asia. (Photo: Courtesy: EPA Reg. 2.)

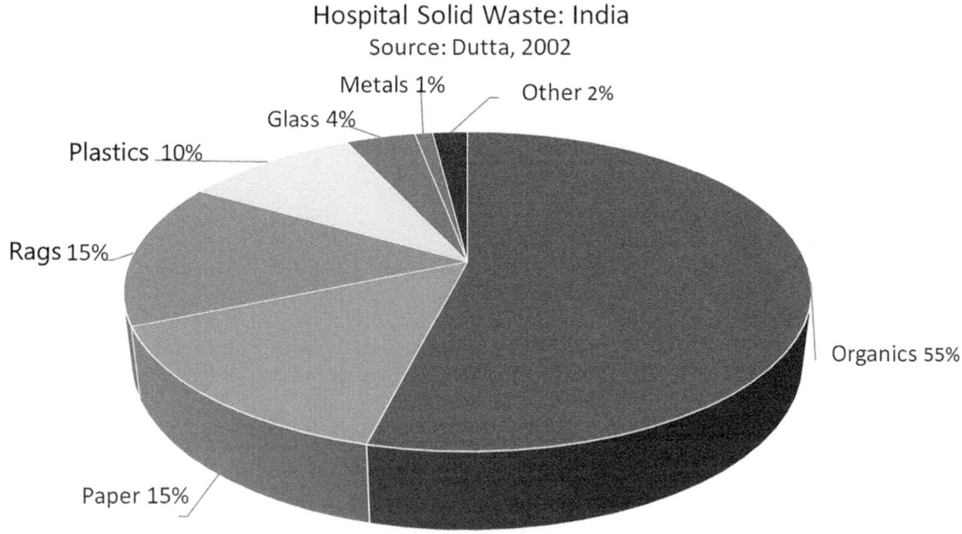

Figure 14.3 Hospital solid waste composition: India.

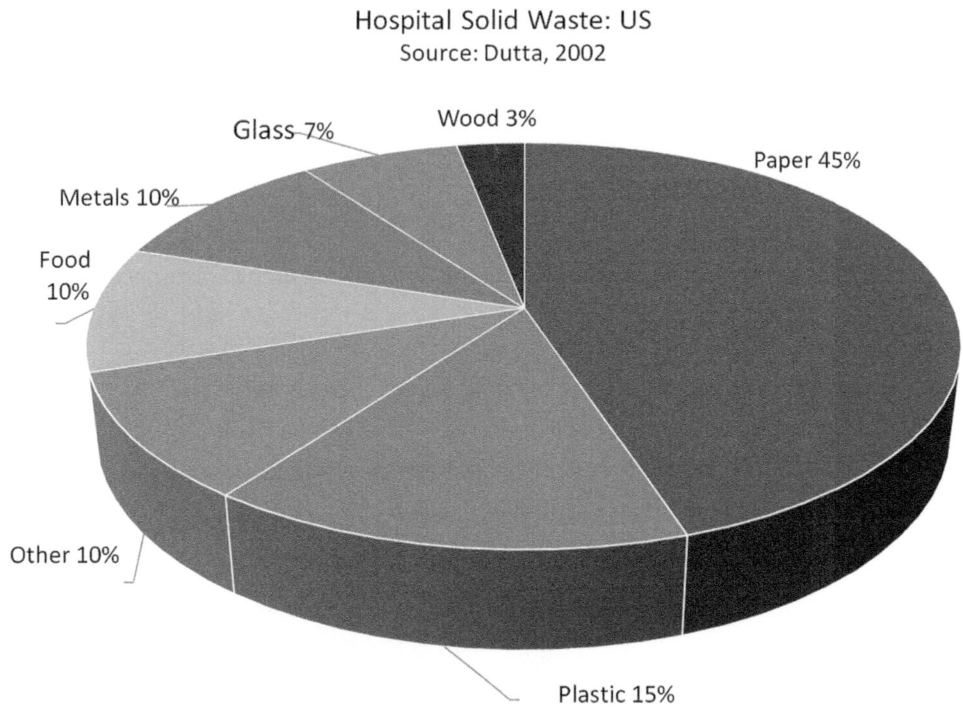

Figure 14.4 Hospital solid waste composition: US.

Table 14.1 Health care waste generation in various regions and countries in the world (WHO, 2014)

Region	*Daily waste generation in kg/bed*
United States of America	7–10
Western	3–6
Latin America	3
Brazil	1–3.2
Eastern Asia	2.5–4
High income countries	2.5–4
Low income countries	1.8 -2.2
Eastern Europe	1.4–2
Eastern Mediterranean	1.3–3

food waste from cafeterias and kitchens. The US hospital's wood component is from pallets and other wood packaging. Hospitals in the US do not report a rag or textile waste stream.

Just as the composition of medical waste may vary among facilities or service providers, between communities or regions, as well as from country to country, the quantities of these waste components also show significant variations. Table 14.1 lists typical waste generation rate for medical facilities in different countries Regions of the World Health Organization (WHO, 2014). The International Healthcare Waste Network has compiled comparative, regional, and available country data on annual hospital waste generation in kilograms per day per occupied hospital bed.

Factors that may influence these variations in waste composition and quantity include the number and specialties of medical facilities, the level of in-patient and out-patient services offered, the proportional use of disposable versus reusable products, the waste prevention methods utilized, the internal storage and waste handling procedures and policies, and access to treatment technologies. The waste generation and medical waste composition in a specific country, region, or community can best be determined by gathering reliable local data.

However, due to the worldwide pandemic, caused by COVID-19, resulting in 172.2 million people infected, and 3.7 million dead, as of June 4, 2021 (daily data from the Johns Hopkins University, Baltimore, MD), the following section of the book covers the waste management aspect of this unanticipated high volume of infectious waste along with some general information on this pandemic.

14.2 WASTE MANAGEMENT FOR COVID-19 PANDEMIC

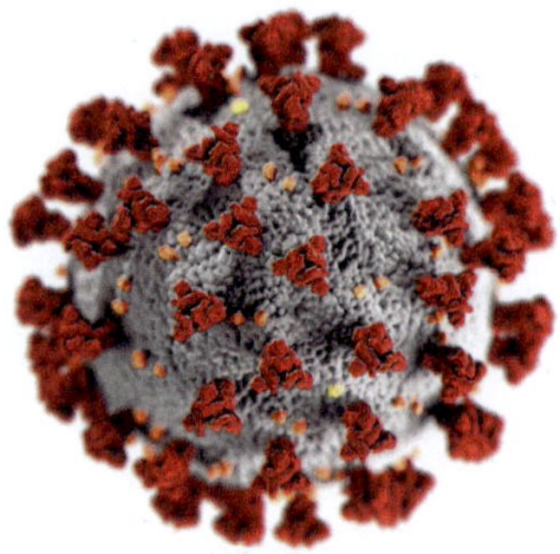

In the recent decade, coronaviruses have caused large-scale pandemics, namely, severe acute respiratory syndrome coronavirus-1 (SARS-Cov-1) and the Middle East respiratory syndrome (MERS). A new outbreak in this family was added in November–December of 2019 as the novel coronavirus disease-2019 (COVID-19) caused by a large group of highly diverse, enveloped, positive-sense, and single-stranded RNA viruses, namely, SARS-CoV-2. An illustration, created at the Centers for Disease Control and Prevention (CDC), reveals ultrastructural morphology exhibited by coronaviruses. Note the spikes that adorn the outer surface of the virus, which impart the look of a corona surrounding the virion, when viewed electron microscopically. The illustration of the SARS-CoV-2 is shown above here (Courtesy of CDC/credits – Alissa Eckert, MSMI; Dan Higgins, MAMS). Soon after the pneumonia disease outbreak in Wuhan (Hubei province, China), the transmission of COVID-19 has been found by human-to-human contact and declared a pandemic of global crisis. Although the genomics sequencing and entry pattern of SARS-CoV-2 into human cells are the same as SARS-CoV-1, the mushroom-shaped spikes (s-) proteins help to maintain the van der Waals forces to keep the RNA intact with the human cellular receptor-ACE2. In addition, the binding affinity of SARS-CoV-2 is 10- to 20-folds higher than SARS-CoV-1. The high affinity of S-proteins for human ACE2 is likely to facilitate the rapid transmission of SARS-CoV-2 and causing the spread of this pandemic situation globally (Ilyas et al., 2020). Currently, mass sampling with rapid tests, isolation of suspected persons, use of personal protective equipment (PPE), surgical and protective facemasks, aprons/gowns, nitrile gloves, and social distancing of six (6) feet or two (2) meters are essentially used as personal protective measures to protect individuals from exposure to this virus and contaminants. These protective measures have been primarily used against pathogens in hospitals as a standard practice under normal conditions without any ongoing crisis. The current pandemic involving COVID-19 has now necessitated the usage of these protective measures in domestic isolation and individual protections, leading to a rapid accumulation of potentially infectious waste streams, causing a challenge with an unpredictably high volume of COVID-19-related waste (C19-waste) in many cities and towns in the world.

This has caused the C19-waste in the US to an estimated increase from 0.5 million to 2.5 million tons/month. This infectious waste also requires treatment without delay to prevent the growth of the pandemic. Since the outbreak in March, South Korea has generated about 2,000 tons of this infectious waste till mid-May, representing an increase of 1,000 tons/month over their routine waste generation. Similarly, Spain generated an unexpected amount of C19-waste with a jump of 350% of their normal medical waste in Catalonia during mid-March 2020 (Ilyas et al., 2020). The drastic increase in a number of countries and people infected with SARS-CoV2 indicates the signs of capacity overrun by the C19-waste. Waste management practices in these countries and many others are seriously strained and are failing to provide safe disposal of these infectious waste, resulting in further spread of COVID-19 to common people and healthcare workers due to these exposures. To that end, it is of high importance and priority to have proper disinfection and disposal practices of C19-waste from the very collection till the final disposition of these infectious wastes.

The minimization, disinfection, and disposal practices of C19-waste are similar to other infectious waste treatment processes as described in the latter part of this chapter. However, to deal with the sudden increase/overload of this infectious waste, some of the interim guidance from the World Health Organization (WHO, 2020) and the US Center for Disease Control (CDC, 2020) are presented in the following subsection to cope with this situation.

14.2.1 Disinfection and disposal of COVID-19-related waste

The disinfection and disposal methods of C19-waste, covered in this subsection, should provide additional guidance to health-care providers, who may be able to glean more information on the water, sanitation, hygiene, and waste-related risks and practices related to C19-waste management. Categorization and classification of medical waste using the methodology described in the latter part of this chapter also include specific aspects of the C19-waste.

The C19-waste needs to be classified at the source or origin to make this most effective in terms of controlling the spread, reducing the waste volume, and providing an appropriate level of disinfection and disposal of the waste. The majority of waste generated in health-care facilities is general, non-infectious waste (e.g., packing, food waste, and disposable hand drying towels). Proper identification, segregation at source should be conducted as follows:

- Segregate general waste from infectious waste and place them in clearly marked bins, bagged and tied, and disposed of as general municipal waste.
- Collect all infectious waste produced during patient care, including C19-waste, safely in clearly marked lined containers and sharp boxes (e.g., sharps, bandages, and pathological waste).
- Treat this waste, preferably on-site, and then dispose safely.

Preferred treatment options are high temperature, dual chamber incineration or autoclaving (WHO, 2020). Other treatment options are provided later in this chapter.

If waste is moved off-site, it is critical to understand where and how it will be treated and disposed. Waste generated in waiting areas of health-care facilities can be classified as non-hazardous and should be disposed in strong black bags and closed completely before collection and disposal by municipal waste services. If such municipal waste services are not available, as interim measure, safely burying or controlled burning may be done until more sustainable and environmentally friendly measures can be put in place. An example of such a small, temporary burial pit for health-care wastes (HCW) is shown in Figure 14.5 (WHO, 1999).

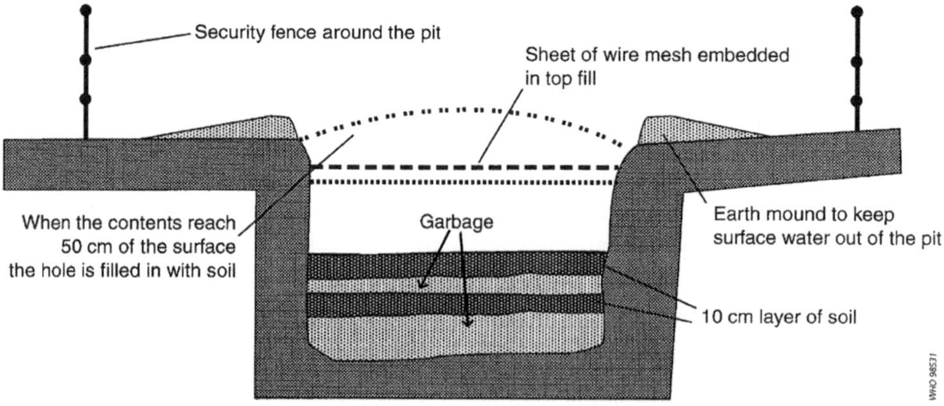

Figure 14.5 Example of a small, temporary burial pit for health-care waste (WHO, 1999).

All those who handle HCW should wear appropriate PPE (long-sleeved gown, heavy-duty gloves, boots, mask, and goggles or a face shield) and perform hand hygiene after removing it. Best practices for safely managing HCW should be followed, including assigning responsibility and sufficient human and material resources to segregate, recycle and dispose of waste safely.

Many cities reported a large increase (five times greater than before the pandemic) of medical waste generated in hospitals, especially through the use of PPE. Therefore, it is important to increase capacity to handle and treat this HCW without delay. Additional waste treatment capacity, preferably through alternative treatment technologies, such as autoclaving or high-temperature burn incinerators, may need to be procured and systems may need to be put in place to ensure their sustained operation. Ideally safe waste disposal is linked to purchasing and investments in PPE. As an interim measure safely burying HCW may be done until more sustainable measures can be put in place. Manual chemical disinfection of waste is not recommended, as it is not regarded as a reliable and efficient method. In addition, countries should work to establish sustainable waste management chains, including addressing logistics, recycling, treatment technologies, and policies (WHO, 2020).

Health-care facilities should follow their existing cleaning and disinfection procedures consistently and correctly. Linens should be laundered and the areas where COVID-19 patients receive care should be cleaned and disinfected frequently (at least twice daily, but more frequently for high touch surfaces such as light switches, bed rails, tables, and mobile carts). Many disinfectants are active against enveloped viruses, such as SARS-CoV-2, including commonly used hospital disinfectants. Currently, WHO recommends (WHO, 2020) using:

- 70% ethyl alcohol to disinfect small surface areas and equipment between uses, such as reusable dedicated equipment (e.g., thermometers);
- Sodium hypochlorite at 0.1% (1,000 ppm) for disinfecting surfaces and 0.5% (5,000 ppm) for disinfection of blood or bodily fluids spills in health-care facilities.

The efficacy of all disinfectants varies with the presence of organic material in the object. Thus, before applying a disinfectant, it is essential to clean surfaces with a detergent and water. The concentration and exposure time of any disinfectant are critical parameters for its efficacy. After applying disinfectant to a surface, it is necessary to wait for the required exposure time and drying to ensure those surface microorganisms are killed.

All individuals in charge of environmental cleaning and laundry of bedding, towels, and clothes from COVID-19 patients should wear appropriate PPE, including heavy-duty gloves, a mask, eye protection (goggles or a face shield), a long-sleeved gown, and boots or closed shoes. They should perform proper hand hygiene after finishing these tasks and after removing PPE. Machine washing with warm water at 60°C–90°C and laundry detergent is recommended. The laundry can then be dried according to routine procedures. If machine washing is not possible, linens can be soaked in hot water and soap in a large drum using a stick to stir, taking care to avoid splashing. The drum should then be emptied, and the linens soaked in 0.05% chlorine for approximately 30 minutes. Finally, the laundry should be rinsed with clean water and the linens dried fully.

WHO also recommends the following protection in handling C19-waste:

- Perform frequent and correct hand hygiene measures to prevent infection with SARS-CoV-2. Proper hand hygiene, using the right technique at the right time either with alcohol-based hand rub or with soap and water, is critical.

- Train sanitation workers on the proper use of PPE and provide access for their use. Water disinfection and wastewater treatment can reduce viruses.
- Use good sanitation, hygiene, and waste management practices to prevent many other infectious diseases as additional health benefits.

The European Agency for Safety and Health at Work has provided general guidance on how to help prevent the spread of the coronavirus at the workplace. Good practices communicated by the European Commission (EC), based upon their stakeholders in the waste management sector, include the following (EC, 2020):

- Adapt proper organization of staff to avoid spreading infection among teams, i.e. respecting distancing between individuals, reducing the number of workers present in the same area to a bare minimum;
- Ensuring the availability and appropriate use of adequate PPE as well as of suitable disinfecting products;
- Ensuring strict adherence to enhanced hygiene standards, including frequent change and cleaning of PPE and professional clothing; replacing professional gloves in the event of breakage or any incident of potential contamination; sanitizing facilities, vehicle cabins, and clothes regularly;
- Ensuring that where masks are usually worn, strict protocols on how to put on and take off PPE are followed, thus avoiding incidental contact and contamination;
- Encouraging specific working conditions, where appropriate, for vulnerable people, such as elderly workers and persons with specific chronic health problems.

The US CDC published a guidance (CDC, 2020) for the waste collectors and recyclers with several recommendations on personal protection measures. CDC recommends wearing cloth face coverings in public settings where other social distancing measures are difficult to maintain, especially in areas of significant community-based transmission. Cloth face coverings may prevent people who don't know they have the virus from transmitting it to others. These face coverings are not surgical masks or respirators and are not appropriate substitutes for them in workplaces where masks or respirators are recommended or required. Some of the CDC recommendations that are unique and specific to certain activities pertaining to waste collection and recycling as listed below:

- Practice routine cleaning and disinfection of frequently touched surfaces, such as steering wheels, door handles, levers, and control panels. Key times for cleaning include:
  - At the beginning and end of every shift and
  - After anyone else uses your vehicle or workstation.
- Wear your normal PPE as you go about your day. This may include work gloves, eye protection (such as safety glasses), and a work uniform or coveralls.
- Use disinfectant to clean eye protection at the beginning and end of your shift.
- Replace work gloves when they become damaged (e.g., if they are ripped or torn). Practice proper hand hygiene, cough and sneeze etiquette. These are important infection control measures. Wash your hands regularly with soap and water for at least 20 seconds or use an alcohol-based hand sanitizer containing at least 60% alcohol. Key times to clean hands include:
  - Before and after work shifts and work breaks
  - After blowing your nose, coughing, or sneezing

  - After using the restroom
  - Before eating or preparing food
  - Before putting on, touching, or removing cloth face coverings.
- Avoid contact with body fluids, if possible. Use gloves if you have to touch surfaces contaminated by body fluids.
- Avoid touching your eyes, nose, or mouth. Be extra careful when putting on or taking off PPE.
- Stay up to date on your company's current policies on COVID-19. Follow the social distancing guidance provided by your employer.

14.2.1.1 Disposal technologies for C19-waste

The COVID-19 waste (C19-waste) collection, handling, and initial disinfection at source have been covered in the earlier sections of this chapter. The waste treatment options are briefly covered here. Further details on various other treatment options are described in Section 14.5.

This waste should be treated, preferably on-site, and then safely disposed. Preferred treatment options are high temperature, dual chamber incineration or autoclaving. If waste is moved off-site, it is critical to understand where and how it will be treated and disposed of. Waste generated in waiting areas of health-care facilities can be classified as non-hazardous and should be disposed of in strong black bags and closed completely before collection and disposal by municipal waste services. If such municipal waste services are not available, as interim measure, safely burying or controlled burning may be done until more sustainable and environmentally friendly measures can be put in place. All those who handle HCW should maintain appropriate safety protocols and gears/clothing as mentioned earlier. (WHO, 2020)

Many cities have reported a large increase (five times or greater than pre-pandemic) in the volume of medical waste generated in hospitals, especially through the use of PPE. Therefore, it is important to increase the capacity to handle and treat this HCW without delay. Additional waste treatment capacity, preferably through alternative treatment technologies, such as autoclaving or high-temperature burn incinerators, may need to be procured and systems may need to be put in place to ensure their sustained operation. Other technologies for treating infectious waste include (WHO, 1999):

• Rotary kiln	• Wet thermal treatment
• Pyrolytic incinerators	• Microwave treatment
• Chemical disinfection	• Encapsulation

Some of the advantages and disadvantages of the above-listed technologies are provided below (WHO, 1999):

The Rotary kiln and the Pyrolytic incinerators are effective for removing all infectious waste, most chemical waste, and pharmaceutical waste. However, they are highly cost-intensive and require the installation of scrubber systems for the treatment of flue gas for proper emission control.

Microwave treatments have good disinfection efficiency under appropriate operating conditions. They generally help with drastic reduction in waste volume and are environmentally sound. Relatively high investment and operating costs. They also come with high operation and maintenance problems.

Chemical disinfection treatment is highly efficient under good operating conditions. Some chemical disinfectants are relatively inexpensive. Causes drastic reduction in waste volume. These systems require highly qualified technicians for the operation process. This technology is inadequate for pharmaceuticals and some chemical and infectious waste.

Wet thermal treatment is environmentally sound. It results in drastic waste reduction. Relatively lower investment and operating cost. However, these systems are inadequate for anatomical, pharmaceutical, chemical, and other wastes that are not readily stream-permeable. They also have shredders which are subject to frequent breakdowns.

Encapsulation is a low cost and safe option for low volume of waste. They can be used as a temporary measure to cope up with a pandemic situation. However, access to the site must be restricted. This option is not recommended for non-sharp infectious waste.

A number of other technologies for the treatment and disposal of infectious waste, including C19-waste, are covered in detail under "Treatment Options for Medical Waste" in Section 14.5.

14.3 MEDICAL WASTE IDENTIFICATION AND PROFILE

14.3.1 Quantum of medical waste

Developing countries are in an advantageous position from the viewpoint of per capita volume of waste generation. Hospital waste generated in developing countries per patient is much less compared to the volume generated in the developed countries. Volume of waste generated from a medical facility in developing countries ranges from 1 to 3 kg/day/bed as compared to 5–8 kg/day/bed in developed countries. Table 14.1, in Section 14.1, provides a list of typical waste generation rate for medical facilities in different countries/regions of the world.

14.3.2 Categorization of medical waste

Medical waste is diversely categorized in many parts of the world since it comprises a variable mixture of wastes. The mixture components include waste from all areas of the facility such as administrative offices, cafeterias, public areas, laboratories (pathology, microbiology, chemical, and research), nuclear medicine, operating rooms, labor rooms, emergency rooms, intensive care units, clinics, and wards. While waste from administrative offices might contain primarily paper, waste from other areas will contain materials such as kitchen garbage, hazardous chemicals, infectious microbes, body parts, or pathology specimens.

These waste materials are very different. Some can be reused or recycled; some are safely disposed of as household waste; some are hazardous and require special (and usually more costly) treatment and disposal techniques to protect the human health and the environment.

Safe and efficient management of waste from a medical facility involves categorizing different waste materials. Waste materials with similar hazard properties and treatment/disposal requirements can be put in the same waste category. National/local regulations frequently specify waste categories and the associated treatment and disposal facilities assigned to each.

For the most part, individual countries use waste category systems that reflect their domestically available treatment and disposal methods and practices. Three major categories of waste pattern are generally found in most countries, which include general waste, infectious waste, and hazardous waste. Individual countries subdivide these major categories in various ways that they find befitting to their specific situation. In the US, infectious waste is categorized as follows:

- Isolation waste
- Cultures and stocks and associated biologicals
- Human blood and blood products
- Pathological waste
- Used sharps
- Contaminated animal carcasses
- Unused sharps.

Various other categorization systems exist in many parts of the globe. Some examples of medical waste classification systems in the Western Pacific countries are provided in Table 14.2. They are all based upon characteristics of waste that require special segregation, handling, and disposal. The categorization of HCW waste by a few Asian Pacific countries, such as Australia, Japan, Malaysia, New Zealand, Papua New Guinea, Philippines, and Singapore is furnished in Table 14.2.

Categorization of medical waste plays a vital role in safe and economical management of the waste. Organizations and regulatory agencies involved in waste management are generally in agreement when it comes to deciding on the specific categories of waste generated by medical facilities. However, instances where they fail to agree on the issue, it creates a problem for the medical waste management community.

Overall, cost of waste handling and disposal is inevitably linked to the type of the waste and their categories as defined. For example, stringently defining the infectious waste category may result in shifting the waste from the general category into the infectious category. Because of the higher costs associated with handling and disposing of the infectious category, overall costs of waste management would thus increase. On the contrary, when fewer waste materials are defined as infectious waste, overall cost would drop, but the risk of infection to people exposed to the waste will increase because of the presence of residual infectious waste in the waste material.

In general, waste generated by medical facilities can be listed into three major categories:

- General waste
- Infectious waste, and
- Hazardous waste.

14.3.2.1 General waste

Most of the waste (75%–90%) generated from a medical facility can be categorized as general waste. This category of waste resembles household waste and originates in offices, corridors, public areas, supply departments, catering areas (other than food), etc. of a medical facility. It includes newspapers, letters, documents, packing materials, cardboard containers, plastic

Table 14.2 Waste categorization adopted by a few Asia Pacific countries (WHO, 1994, 2015)

Country	*Waste categories*	*Country*	*Waste categories*
Australia (National)	• Sharps • Infectious waste • Human tissues • Cytotoxic • Pharmaceutical • Chemical • Radioactive • Plastic	China	• Infectious waste • Pathological waste • Sharps • Pharmaceutical waste • Chemical waste
Japan	• Infectious waste • Blood and body fluids • Pathological • Sharps • Laboratory/ Culture • Other infected materials	Malaysia	Clinical waste: • Infectious • Sharps • Laboratory • Pharmaceutical and cytotoxic Others: • Radioactive waste • Chemical waste • Pressurized containers • General waste
Mongolia	Hazardous waste • Highly Infectious • Infectious • Sharps • Pharmaceuticals • Chemical • Cytotoxic • Radioactive • Pathological • Pressurized/metal containers Non-hazardous- household waste	New Zealand	General waste • Ordinary waste • Kitchen waste Special waste • Anatomical • Soiled dressing • Infectious waste • Disposables • Laboratory waste • Chemical and pharmaceutical Radioactive waste Cytotoxic waste
Papua New Guinea	Liquid waste Solid waste • Special hospital waste • Trash • Radioactive waste	Philippines	• Dry waste • Wet waste • Sharps • Clinical waste • Pathological waste • Chemical waste • Radioactive waste
Singapore	• Infectious waste • Pathological waste • Routine clinical • Contaminated sharps • Cytotoxic • Radioactive • Pharmaceutical • Chemical waste • General waste		

bags/films, food wrappings, metal cans, food containers, flowers, floor sweepings, and kitchen waste. General waste may be further sorted for partial recycle/reuse purposes and the rest disposed of as municipal solid waste.

14.3.2.2 Infectious waste

About 15% of a medical facility's waste would typically be categorized as infectious waste. Infectious waste originates in many hospital departments, wards, and laboratories and may be defined and/or subject to national/local regulations.

Infectious waste is known by many different names: infectious, pathological, biomedical, biohazardous, toxic, or medically hazardous waste. Although its definition has been debated for years by regulatory agencies, hospitals, and research laboratories, there is still no universally accepted definition for infectious waste. The single characteristic common to all infectious waste is its ability to infect. The infection process is not a simple one. It requires the interaction of at least the following four factors:

- dose
- presence of a pathogen of sufficient virulence
- portal of entry
- resistance of host.

Given the complexity of these four factors and their interactions, it is not surprising that there are no widely accepted objective tests for determining whether a particular waste is infectious. Thus, designating a waste or waste stream as infectious usually rests on professional judgment rather than on the results of objective tests.

As shown in Table 14.2, regulatory agencies of various countries tend to agree on the categorization of waste from a medical facility. Table 14.3 indicates the general consensus between the United States Environmental Protection Agency (EPA), the United States Centers for Disease Control (CDC), and the WHO/Western Pacific Region (WPR) concerning the categorization of infectious waste.

Table 14.3 Variations in classification of infectious waste

Type of waste	*Classification of infectious waste*		
	EPA[a]	CDC[a]	WHO/WPR[a]
Biological wastes	Yes	Yes	Yes
Pathological wastes	Yes	Yes	No[b]
Blood and blood products	Yes	Yes	Yes
Contaminated animal carcasses	Yes	No	Yes
Isolation wastes	Yes	Subject to facility Policy	Yes
Used sharps	Yes	Yes	No[b]
Some unused sharps	Yes	No	No

[a] EPA – USA Environmental Protection Agency; CDC - USA Centers for Disease Control; WHO/WPR - World Health Organization, Western Pacific Region.

[b] WHO/WPR classifies pathological waste and used sharps as "hazardous health-care waste" rather than as infectious waste.

The EPA has defined infectious waste as waste, which is capable of producing an infectious disease. More practically, and for regulatory purposes, EPA has designated seven waste types, listed in Section 14.3.2, under the infectious waste category.

The CDC looks at infectious waste from a public health viewpoint and has defined it in different terms. According to CDC, infectious waste includes only blood and other body fluids containing visible blood; semen and vaginal secretions; and other specified fluids. CDC further notes that it designated these wastes not as infectious, *per se*, but simply as those that require health-care workers to adhere to "universal precautions".

The WHO from its international public health viewpoint, defines infectious waste as "waste suspected to contain pathogens" and cites (1) laboratory cultures, (2) waste from isolation wards, (3) tissues, materials, or equipment which has been in contact with infected patients, and (4) excreta as examples. Notably, WHO classifies both pathological waste and sharps as "hazardous health-care" waste rather than infectious waste.

14.3.2 3 Hazardous waste

Medical facilities use many hazardous chemicals in both the diagnosis and the treatment of patients, as well as in general operation of the facility. Hazardous chemicals such as the ones listed below would typically be found in a medical facility.

- cytotoxic chemicals (chemotherapy and antineoplastic chemicals)
- formaldehyde
- photographic chemicals
- radionuclides
- solvents
- mercury
- anesthetic gases
- cleaning chemicals
- maintenance chemicals.

Hazardous waste from the use of these chemicals originates from a variety of sources within the medical facilities. It occurs primarily as waste from clinical laboratories and associated services. Chemical waste also includes discarded solids, liquid and gaseous chemicals from diagnostic and experimental work, cleaning, housekeeping, and disinfecting procedures. Hazardous chemical waste generated by a medical facility, as compared to an industrial facility, is unique in that it contains a broad range of chemicals but in small volumes, whereas in the case of latter, it is just the opposite. There may be national/local regulations which apply to hazardous chemical waste.

Cytotoxic drugs, also known as antineoplastic drugs or cancer chemotherapy drugs, have the ability to kill or arrest the growth of living cells. They play an important part in the therapy of various neoplastic conditions but are also finding a wider role as immuno-suppressive drugs in transplantation and in treating various diseases with an immunological basis. Cytotoxins are commonly administered by injection or infusion but some are given by mouth in tablet, capsule, or suspension form. Cytotoxic drugs are most often used in specialist units such as oncology and radiotherapy units whose main work is the treatment of cancer. However, there is a significant and increasing usage in other wards and departments of hospitals as well.

Animal studies have shown that at high doses, many of these substances are carcinogenic and mutagenic. Some are also teratogenic. Many cytotoxic drugs are extremely irritating, producing harmful local effects after direct contact with the skin or the eyes. There are three possible routes of entry to the body:

- *Inhalation*: inhalation may occur where an aerosol or airborne dust is generated due to a poor technique or accidentally during handling.
- *Ingestion*: ingestion is unlikely to occur if good hygiene procedures are followed, as should be the case in the handling of any drug by professional personnel
- *Skin contact*: apart from purely irritant effects, certain elements of these drugs may be absorbed into the body through the skin if not removed by washing.

The main opportunity for exposure to health-care personnel arises during the preparation and manipulation of injectable cytotoxins, which may be presented as freeze-dried material or powder, requiring it to be mixed with a diluent. Exposure to cytotoxic material is also possible in other ways, particularly if capsules are opened to remove the powder they contain.

Cytotoxic waste includes expired cytotoxic drugs and materials, such as swabs, tubing, towels, and sharps contaminated with cytotoxic substances during the preparation, transportation and administration of cytotoxic drug therapy.

Formaldehyde is used in general nursing, pathology, autopsy, dialysis, embalming and preservation of specimens. Formaldehyde is a significant source of chemical waste at many medical facilities. Solutions from formaldehyde waste should be safely managed and not discharged into the sewer. Discharging a hazardous material into the sewer is an undesirable management practice. Proper handling and management of hazardous medical waste is described in the following sections.

Photographic chemical developing solutions are used in X-ray departments and consist of two parts: a fixer and a developer. The fixer normally contains 5%–10% hydroquinone, 1%–5% potassium hydroxide, and less than 1% silver. The developer contains approximately 45% glutaraldehyde. Stop baths and fixer solutions contain acetic acid. Silver is usually recovered from spent fixer solutions with the remainder typically being discharged to the sewer.

Radionuclides used as radioactive substances in medical facilities are either “open sources”, which involve the direct use of a radiochemical substance, or “closed sources”, involving the indirect use of a radiochemical substance enclosed in a piece of equipment.

Open sources are chemicals which possess low levels of radioactivity and give rise to low-level radioactivity. A low-level radioactive waste is under 1 MBq (megaBecquerel) or 0.0233 mCi (millicurie). Most of this waste is generated during organ imaging and tumor localization, with a radioisotope for each procedure in the MBq activity range. Radioactive waste of lower activity may be produced during in vitro (conducted outside of any living organism) diagnostic studies. However, the therapeutic application of radioiodine, while being infrequent, is of a much higher level of activity, being of the order of 1 GBq (gigaBcquerel), and produces waste with a significantly higher level of radioactivity.

“Sealed sources” use radionuclides possessing high levels of activity. These sources do not routinely generate radioactive waste. The most common method of dealing with these sources is to return them to the supplier when exhausted or when no longer required.

Medical facilities have three main sources of radioactive waste:

- research activities, which commonly use significant quantities of 14-C and 3-H and generate large volumes of waste with low radioactivity
- clinical laboratories, which are involved in radioimmunoassay procedures that likewise generate relatively large volumes of waste with low radioactivity
- nuclear medicine laboratories, which will normally generate relatively small amounts of waste but with higher radioactivity than the previous two sources.

Radioactive waste includes solid, liquid, and gaseous waste contaminated with radionuclides generated from "in vitro" analysis of body tissues and fluid, "in vivo" (conducted inside of living organism) body organ imaging and tumor localization, and therapeutic procedures. Examples include:

- *Solid*: vials, syringes, glassware, absorbent paper, swabs
- *Liquid*: residues from shipments of radioactive material, unwanted solutions of radionuclides intended for therapeutic use, patient's urine (and excreta)
- *Gas*: clinical application of ^{85}Kr and ^{133}Xe: exhausts from stores (e.g., radium stores) and fume cupboards.

Radioactive waste from medical facilities is usually stored until the level of radioactivity has decayed to a safe level and then disposed of safely in a manner consistent with the chemical properties of the material.

Solvent wastes are typically generated in various departments throughout a hospital. These include pathology, histology, engineering, and laboratories. Specific solvents used include halogenated compounds such as methylene chloride, chloroform, freons, trichloroethylene, and 1,1,1-trichloromethane. Other solvents include non-halogenated compounds such as xylene, acetone, methanol, ethanol, isopropanol, toluene, ethyl acetate, and acetonitrile. Acetone, methanol, and ethanol wastes are usually evaporated or discharged to the sewer; other solvent wastes are normally handled as hazardous wastes. Disposal methods may involve solvent recovery, incineration with or without heat recovery or landfilling.

Mercury wastes are primarily generated at most hospitals from broken or obsolete equipment. Mercury wastes are decreasing in quantity because solid-state electronic sensing instruments (thermometers, blood pressure gauges, etc.) are replacing those containing mercury.

Mercury spilled from broken equipment can be recovered and reused (if uncontaminated), but this is not frequently done. Details on safe management of mercury are provided in the following sections.

Waste anesthetic gases, such as nitrous oxide, the halogenated agents halothane (Fluothane), enflurane (Enthrane), isoflurane (Forane), and other substances are used as inhalation anesthetics. Exposure of health-care personnel to these substances may result in acute toxic effects and, possible reproductive disorders and carcinogenesis.

These wastes may cause acute toxic effects on health-care personnel. Halogenated compounds are supplied in liquid form, generally in glass bottles. Once empty, the bottles could be reused (sent back to the supplier for refill).

Toxics, Corrosives, and Miscellaneous Chemicals include poisons, oxidizers, and caustics that are used throughout most hospitals, generally in small quantities. Waste oils and solvents from maintenance activities may also be classified as hazardous wastes as they exist in some boiler water conditioning chemicals. The following list describes major toxic, corrosive, and miscellaneous chemical wastes from medical facilities and their sources of origin:

- *Ethylene oxide*: used in sterilizers. Classified as a probable human carcinogen; also a smog-forming agent, and an explosive/flammability hazard.
- *Disinfecting cleaning solutions*: phenol-based, used for scrubbing floors and other applications
- *Maintenance wastes*: waste lube oils, vacuum pump oils, cleaning solvents, paint stripping wastes, leftover paints and painting accessories, and spent fluorescent lamps.
- *Utility wastes*: boiler feedwater treatment residuals (resin regeneration brine, spent resin), boiler blowdown, boiler cleaning (lay-up) wastes, cooling tower blowdown, and cooling tower sludges/sediments.

14.3.3 Impact of waste categorization on waste management

Infectious waste categorization is generally agreed upon by various organizations and regulatory agencies as mentioned earlier. This categorization is used to classify the medical waste by these organizations and agencies. However, where there are disagreements in categorization between regulatory organizations, it causes problems and confusion for the medical waste facilities in terms of regulatory compliance in those concerned areas.

The overall cost of waste handling and disposal is inevitably tied to how wastes are defined. While considering the interaction between the general and infectious waste categories, a precise definition of the infectious waste may change the classification of general category waste into the infectious category. Consequently, the higher costs associated with the handling and disposal of the infectious waste category will lead to an increase in the overall costs. On the contrary, when fewer waste materials are defined as infectious, the overall cost would decrease.

Some medical facilities, mostly in developing countries, question whether they can reliably segregate infectious and general waste in certain areas of their facility. As a result, they arbitrarily define all wastes from such areas as infectious to minimize any chances of crossover of infectious waste into the general waste stream. Thus, to avoid worker safety and environmental liability, they intentionally dispose of some of their general waste as infectious waste, thereby incurring much higher costs. Although the increase in cost of such practices is passed on to the patients, the ultimate impact is much greater in terms of energy usage and abuse of valuable landfill space.

Similarly, conflicting definitions confuse everybody and generally result in a facility decision to follow the more strict definition as a strategy for avoiding liability.

Redefining waste categories and consequently shifting waste materials from one category to another can cause bottlenecks in what was previously a smoothly operating waste handling, treatment and disposal system. Specific problems might include inadequate collection frequencies, excessive accumulations in storage rooms, and backups of waste collection carts in corridors and treatment areas.

14.4 MEDICAL WASTE MINIMIZATION OPTIONS

Key strategies to minimize medical wastes include:

- Segregation of individual waste streams
- Keeping hazardous waste segregated from non-hazardous waste

- Minimizing dilution of hazardous wastes, and
- Clearly marking all chemical and waste containers.

14.4.1 Management and control practices

Use of the following management and control practices is suggested for minimizing medical waste:

- Centralize purchasing and dispensing of drugs and hazardous chemicals
- Track/monitor drug and chemical flow within the facility
- Conduct periodic waste audits of each department generating wastes
- Apportion waste management costs
- Improve inventory control (e.g. require use up of old stocks before ordering new stocks of chemicals; and order hazardous chemicals only when needed)
- Providing employee training on:
  - o Spill prevention and preventive maintenance
  - o Emergency preparedness and response to spill cleanup
- Implement facility-wide waste reduction program.

14.5 TREATMENT OPTIONS FOR MEDICAL WASTES

A lot of research has gone into the development of suitable treatment technologies for a safe and effective management of medical waste (USEPA, 1994). New processes and technologies are being introduced rapidly to meet the needs and demands of the medical facilities.

Medical waste treatment systems are designed and operated to achieve the basic requirements of decontamination, revitalization, sterilization, or destruction of the waste/used products according to the desired end use.

The following are the different types of treatment technologies commonly used for treating medical wastes.

14.5.1 Short wave radio frequency (RF) treatment

The dielectric heating or the radio frequency (RF) heating treatment system has been described earlier in Chapter 10, Section 10.2. In this application also, the RF heating technique is used for treating the infectious medical waste. The process involves exposure of shredded medical wastes material to high strength, low frequency (short wave radio frequency) radiation to heat the waste to the desired temperature (>90°C). The heated waste is then stored in the insulated containers in which it is treated to maintain the elevated temperature for 4 hours. At the end of the storage period, the waste is disposed of in a landfill, or recycled as refuse-derived fuel, or the segregated plastic portion of the waste may be sold as recycled material.

The following waste types are considered suitable for the RF heating treatment (USEPA, 2012; Dutta, 2002):

- Cultures and Stocks of infectious agents and associated biologicals, including cultures from medical and pathological laboratories.

- Cultures and stocks from infectious agents from research and industrial laboratories.
- Waste from the production of biologicals, discarded live and attenuated vaccines,
- Culture dishes and devices used to transfer, inoculate,
- Pathological wastes, human blood and blood products.
- Intravenous bags, used sharps.
- Other types of broken or unbroken glassware, which were in contact with infectious agents, such as used slides and coverslips.
- Contaminated Animal carcasses and body parts.
- Isolation wastes, and
- Unused sharps.

14.5.2 Chemical/mechanical treatment

A chemical/mechanical system uses antimicrobial chemicals alone or in combination with a mechanical shredder/hammer mill to increase the contact area for destroying infectious and other harmful microorganisms. Several levels of antimicrobial activity are defined to indicate the types of organisms the chemical agent class is expected to kill.

The effectiveness of the treatment depends upon the characteristics of the chemical agent, the concentration of active ingredient, the contact time with the waste, and the waste characteristics.

In this treatment system, jars of liquid antimicrobial chemicals are often used in small clinics and medical facilities. The solution is placed in a large container where the infectious waste is dumped during the working hours of the facility. In this type of treatment, periodic addition of the chemical agent is needed to maintain the strength of the disinfectant. At the end of the day, the disinfected liquid may be disposed of in the sanitary sewer system, and all disinfected solid waste should be disposed of in the municipal solid waste disposal system, if permissible by the existing regulations.

The following waste types are considered suitable for the Chemical/Mechanical treatment:

- Cultures and stocks of infectious agents and associated biologicals, including cultures from medical and pathological laboratories.
- Cultures and stocks from infectious agents from research and industrial laboratories.
- Waste from the production of biologicals, discarded live and attenuated vaccines.
- Culture dishes and devices used to transfer and inoculate pathological wastes
- Human blood and blood products. Intravenous bags are also included in this category.
- Used sharps. Also included are other types of broken or unbroken glassware that were in contact with infectious agents, such as used slides and coverslips. Waste containing broken glass should be handled with special care and suitably containerized to avoid problems and to provide protective handling of the waste.
- Contaminated bedding of animals, and items used in research activities that have been exposed to infectious agents could be treated by this technology. However, this technology is not suitable for treating contaminated animal carcasses and body parts.
- Isolation wastes, and
- Unused Sharps.

14.5.3 Steam autoclave treatment

Various types of steam autoclave treatment systems are used for treating medical waste. Steam autoclaving combines moisture, heat, and pressure to inactivate microorganisms. This process has been in use for many years and the validity of this sterilization technique is well documented and established (WHO, 2019; USEPA, 1993).

An autoclave consists of a metal vessel designed to withstand high pressures, with a sealable door and an arrangement of pipes and valves through which steam is introduced into and removed from the vessel. Because air is an effective insulator and a key factor in determining the efficiency of steam treatment, removal of air from the autoclave is essential to ensure penetration of heat into the waste. Waste treatment autoclaves must also treat the air removed at the start of the process to prevent pathogenic aerosols from being released. This is usually done by treating the air with steam or passing it through a specific filter (e.g., High Efficiency Particulate Air (HEPA) filter or microbiological filter) before being released. The resulting condensate must also be decontaminated before being released into the wastewater system.

Autoclaves are generally constructed with a metal chamber to withstand the increased pressure/temperature required to insure the destruction of bacteria, viruses, and bacterial spores. Autoclaves come in two basic varieties:

- Gravity displacement autoclaves and
- Pre-vacuum autoclaves.

The size of the device may vary from small, bench-top models to large commercial devices. The bench-top models are generally designed to hold a single bag of waste and the large commercial devices can treat more than a ton of waste per cycle. A typical autoclave system setup in a medical waste facility in Asia is shown in Figure 14.6.

Gravity displacement autoclave: The Gravity displacement autoclave relies on the convective movement of airstream between the hot/lighter and the cold/heavier air. The airstream enters at the top of the device and gradually replaces the existing cooler air as it moves

Figure 14.6 Typical autoclave system in a medical facility. (Courtesy: EPA Region 2.)

towards the outlet at the bottom of the chamber. The efficiency of the system depends on the method of packing and loading of the waste into the autoclave to prevent the formation of air pockets where the existing air may not be displaced by the steam. The operating temperature of the gravity displacement autoclave may be lower than that in pre-vacuum autoclaves and the resulting steam penetration may be less complete in this type of system.

Pre-vacuum autoclave: The pre-vacuum autoclaves create a high vacuum in the treatment chamber by removing air from the chamber before the introduction of steam. This procedure allows the autoclave to reach operating temperatures more rapidly and allows the steam to penetrate the entire load more completely by reducing the chances for concealed air pockets within the waste load. Commercially available pre-vacuum autoclaves are generally designed to operate at about 132°C for the treatment of medical waste.

The following waste types are considered suitable for the Steam Autoclave treatment:

- Cultures and stocks of infectious agents and associated biologicals, including cultures from medical and pathological laboratories.
- Cultures and stocks from infectious agents from research and industrial laboratories.
- Waste from the production of biologicals, discarded live and attenuated vaccines.
- Culture dishes and devices used to transfer or inoculate pathological wastes.
- Human blood and blood products including Intravenous bags.
- Used sharps and other types of broken or unbroken glassware, which were in contact with infectious agents, such as used slides and coverslips. Waste containing broken glass should be handled with special care and suitably containerized to avoid problems and to provide protective handling of the waste.
- Contaminated bedding of animals, and items used in research activities that have been exposed to infectious agents.
- Isolation Wastes.
- Unused sharps.

Contaminated animal carcasses and body parts are not suitable for treatment by the Steam Autoclave System.

14.5.4 Microwave treatment

Microwave treatment of medical waste is a common practice in various parts of the globe where medical facilities and clinics have access to stable supply of electricity. Two types of microwave treatment systems are generally used. A small, unsophisticated, bench-top microwave oven is generally used on an on-site basis by small clinical facilities and research laboratories. Commercially available, large industrial microwave ovens, which are generally used by large hospitals, and commercial waste treatment facilities for treating large quantities of waste.

A typical microwave treatment system generally

- Uses mechanical grinder/shredder
- Sprays about 10% moisture
- Uses microwave heat to keep waste at ≥90°C for 2 hours.

The following waste types are considered suitable for the microwave treatment:

- Cultures and Stocks of infectious agents and associated biologicals, including cultures from medical and pathological laboratories.
- Cultures and stocks from infectious agents from research and industrial laboratories.
- Waste from the production of biologicals, discarded live and attenuated vaccines.
- Culture dishes and devices used to transfer or inoculate pathological wastes.
- Human blood and blood products, that are present as stains and in dried form including intravenous bags.
- Used sharps and other types of broken or unbroken glassware, which were in contact with infectious agents, such as used slides and coverslips. Waste containing broken glass should be handled with special care and suitably containerized to avoid problems and to provide protective handling of the waste.
- Contaminated bedding of animals, and items used in research activities that have been exposed to infectious agents could be treated by this technology. However, this technology is not suitable for treating contaminated animal carcasses and body parts.
- Isolation wastes, and
- Unused Sharps. Only ground, shredded or otherwise destroyed metal items (e.g., hypodermic needles, special blades) are suitable for treatment in a microwave treatment system (Dutta, 2002, USEPA, 1993).

Discarded organs and body parts are not suitable for treatment by the microwave treatment system. Also, bulk metal materials are not suitable for treatment by the microwave treatment system.

14.5.5 Incineration treatment

The Incineration treatment is generally used very commonly in destructive treatment of medical wastes. Many different types of incineration technologies are used for destroying the contaminants from medical waste (Freeman and Harris, 1995). The Incineration treatment technology is covered in detail under Chapter 9 of this book.

Readers are referred to Chapter 9 of this Book for details on *Incineration Treatment* systems including the Incineration Figures referred in this chapter. A case study presented in Chapter 9 includes various information on incineration treatment, operational parameters, emissions, and many other responses to frequently asked questions (FAQs) involving community/public concern for installation of an Incineration Treatment System at the Washington State University, in the state of Washington, west coast of USA.

The following treatment technologies and processes are listed as a few examples of incineration treatment:

- Flame oxidation
- Controlled chamber combustion
- Catalytic oxidation

- Plasma arc and Infrared incineration
- Liquid injection incinerators
- Fixed/open hearth incinerators
- Rotary kiln incinerators
- Fluidized bed incinerators
- Gas or fume incinerators.

All of the incinerator figures referred here are included in Chapter 9. A schematic diagram of a typical incineration facility is shown in Figure 9.1; a liquid injection incinerator is shown in Figure 9.2; details of a typical fixed/sloped hearth incinerator are given in Figure 9.3; and a typical multiple hearth and a rotary kiln incinerators are shown in Figures 9.4 and 9.5, respectively.

Most waste types are considered suitable for the Incineration treatment. However, some organic waste containing polyvinyl chloride (PVCs) and other plastic materials are considered unsuitable because of the possibilities of emissions of dioxins and furans from these waste products. Some of the limitations of incineration treatment, pertaining to the treatment of medical waste are furnished below (USEPA, 1997):

- PVCs and other Plastic materials are not suitable for incineration because of their high potential for emitting products which are highly toxic and persistent, such as dioxins and furans.
- Since incineration does not destroy most inorganic (metals) wastes, this treatment technology may not be effective for waste media containing metals.
- Incineration could be very fuel-intensive for wastes with high moisture content.
- Wastes having low organic content could be costly to incinerate.
- Some wastes containing explosive materials may require a specially designed incinerator.
- Incineration will not be applicable to wastes requiring in situ treatment.
- Wastes with high debris/large particle content may pose problem for some incinerators.

14.6 EVALUATION AND SELECTION OF TREATMENT OPTIONS FOR MEDICAL WASTES

This section is purposely written in a point/bullet format for the readers, who are presumed to have gained knowledge about the technical details from the previous parts of this book. However, if the reader needs any further details on the issues that are unclear, they should consult Chapters 4–10 of this book and the cited literature.

14.6.1 Key factors for effective treatment of medical waste

- Segregation of the waste to get the proper waste to the treatment
- Source separation and segregation using separate containers
- May exclude certain wastes from sterilization treatment in autoclaves
- Separation/segregation of liquids for incineration/other treatment
- Separation/segregation of PVC materials from incineration feed
- Minimize plastics and glasses in incineration feed.

14.6.1.1 On-site vs. off-site treatment considerations

A. Availability of Equipment on-site and Ease of Installation
   - On-site – Advantages :
     - Lower cost,
     - Control over waste, and
     - No transportation need.
   - On-site – Disadvantages:
     - Lack of trained personnel,
     - Improper operation,
     - Additional cost for infrastructure, and
     - Regulatory needs related to waste operations.
   - Off-site – Advantages:
     - Less infrastructure/personnel need for waste operations,
     - No need to check operational conditions,
     - Reap benefit of large treatment facility off-site,
     - Focus better on primary job,
     - Pool resources to get better emission control equipment.
   - Off Site – Disadvantages:
   - Transportation and packaging,
   - Lack of control on waste, and
   - Possibility of future liability due to improper treatment.

B. Regulatory Requirements
   - Technology selected must meet all regulatory requirements.
   - Permits, wastewater treatment/disposal needs.

C. Ease/difficulty of Operation
   - Availability of trained personnel for proper operation.
   - Availability of spare parts/repair support.
   - Additional security/infrastructure need for the equipment.

D. Treatment Applicability
   - Ease/difficulty of operation.
   - Importance of waste separation.
   - Separation/segregation of certain items may be necessary for effective treatment.

E. Effects of Treatment on Waste
   - Appearance of waste after treatment.
   - Volume reduction.

F. Occupational Hazards
   - Dangers of burns or cuts to operators.
   - Exposure to toxic vapors during loading/unloading.
   - High noise level and injury from moving parts.

G. Environmental Effects
   - Emissions from incinerators.
   - Wastewater from steam sterilizers.
   - Solid residues/ashes with possible high metals contents.

H. Costs
   - Capital cost.
   - Operation & maintenance (O&M) costs.

- Cross-media transfer control costs.
- Monitoring costs.
- Operator/personnel costs.

I. Reliability
- Equipment failure/downtime.
- Period of downtime for maintenance/repair.

For emission Controls, BMPs and Technology Comparison, *see* previous chapters, especially Chapters 9 and 10.

REFERENCES

Centers for Disease Control (CDC). (2020). What Waste Collectors and Recyclers Need to Know about COVID-19, Washington, DC, May. [Online] Available from: https://www.cdc.gov/coronavirus/2019-ncov/community/organizations/waste-collection-recycling-workers-h.pdf [Accessed 24th September, 2020].

Dutta, S. (1998). Best Practicable Management of Medical Waste in India. *Proceedings, National Workshop on Management of Hospital waste, sponsored by the Indian Institute for Rural Development (IIRD)*, April 16–18, 1998, Jaipur, India.

Dutta, S. (2002). *Environmental Treatment Technologies for Hazardous and Medical Wastes, Remedial Scope and Efficacy.* Tata McGraw Hill Publishing Company, New Delhi.

European Commission (EC). (2020). Waste Management in the Context of the Coronavirus Crisis, April. 2020 European Commission. [Online] Available from: https://osha.europa.eu/en/highlights/-covid-19-guidance-workplace [Accessed 23rd September, 2020].

Freeman, H. M., and E. F. Harris. (1995). *Hazardous Waste Remediation: Innovative Treatment Technologies.* Technomic Publishing Co., Inc., Lancaster, PA.

Ilyas, S., R. R. Srivastava, and H. Kim (2020). Disinfection Technology and Strategies for COVID-19 Hospital and Bio-medical Waste Management, Elsevier Public health Emergency Collection. [Online] Available from: https://www.ncbi.nlm.nih.gov/pmc/articles/PMC7419320/, August 2020 [Accessed 4th June, 2021].

Kwok-Kuen, D. (1998). *Personal Communication with USAEP.* Environmental Protection Dept., Hong Kong, March.

Pathak, S., (1998). *Management of Hospital Waste: A Jaipur Scenario.* National Workshop on Management of Hospital Waste, Jaipur, April 1998, p. 31.

USEPA. (1990). *Guides to Pollution Prevention – Selected Hospital Waste Streams.* Washington, DC. EPA-625-7-90-009.

USEPA. (1993). *Evaluation of Medical Waste Treatment Technologies, Final Report.* Office of Solid Waste, Washington, DC, January.

USEPA. (1994). *Remediation Technologies Screening Matrix and Reference Guide.* West Chester, PA. EPA/542/B-94/013, October.

USEPA. (1997). *Best Management Practices (BMPs) for Soil Treatment Technologies.* Washington, DC. EPA-530-R-97-007, May.

USEPA. (2012). *A Citizen's Guide to In situ Thermal Treatment.* Washington, DC. EPA 542-F-12-013, September.

World Health Organization (WHO). (1994). *Draft Regional Guidelines for Health Care Waste Management in Developing Countries.* WHO, Western Pacific Regional EHC, Malaysia.

WHO. (2014). *Safe Management of Wastes from Health-care Activities*, edited by Y. Chartier et al. 2nd Ed. WHO, Geneva.

WHO. (2015). *Status of Health-Care Waste Management in Selected Countries of the Western Pacific Region, 2008–2013.* WHO, Geneva.

WHO. (2018). Health-Care Waste – Fact Sheet. Geneva, February. [Online] Available from: https://www.who.int/news-room/fact-sheets/detail/health-care-waste [Accessed 21st September, 2020].
WHO. (2020). *Water, Sanitation, Hygiene, and Waste Management for SARS-CoV-2, the Virus that Causes COVID-1: Interim Guidance*. WHO and UNICEF, Geneva. Reference number: WHO/2019-nCoV/IPC_WASH/2020.4, July.

Subject Index

Note: Bold page numbers refers to title, *italic* page numbers refers to figures, and underline page numbers refers to lists.

For Product Safety Concerns and Information please contact our EU representative GPSR@taylorandfrancis.com Taylor & Francis Verlag GmbH, Kaufingerstraße 24, 80331 München, Germany